Sneeze

David Miles is an infectious disease immunologist who spent ten years researching diseases of childhood in Africa and the vaccinations that protect against them. He now lives in London and tutors on the London School of Hygiene and Tropical Medicine's online MSc course.

SNEEZE

DAVID MILES

First published in the United Kingdom in 2026 by

August Books, an imprint of
Canelo Digital Publishing Limited,
20 Vauxhall Bridge Road,
London SW1V 2SA
United Kingdom

A Penguin Random House Company
The authorised representative in the EEA is Dorling Kindersley Verlag GmbH. Arnulfstr. 124, 80636 Munich, Germany

Copyright © David Miles 2026

The moral right of David Miles to be identified as the creator of this work has been asserted in accordance with the Copyright, Designs and Patents Act, 1988.
All rights reserved. No part of this publication may be reproduced or transmitted in any form or by any means, electronic or mechanical, including photocopy, recording, or any information storage and retrieval system, without permission in writing from the publisher.
No part of this book may be used or reproduced in any manner for the purpose of training artificial intelligence technologies or systems. In accordance with Article 4(3) of the DSM Directive 2019/790, Canelo expressly reserves this work from the text and data mining exception.

A CIP catalogue record for this book is available from the British Library.

Ebook ISBN 978 1 80436 925 8
Hardback ISBN 978 1 80436 924 1

Illustrations by Sam Andebonn

Printed and bound in Great Britain by Clays Ltd, Elcograf S.p.A.

Look for more great books at
www.augustbooks.co | www.dk.com

To all the research volunteers, laboratory technicians, postgrads, postdocs, research nurses, data managers, statisticians, medical officers, field workers and everyone else who contributed to the science described in this book but is not mentioned in it.

Every scientist whose name appears in these pages is standing on your shoulders.

Contents

Introduction	ix
Section 1: How colds work	1
Chapter 1: What is a cold?	3
Chapter 2: When did the first person catch the first cold?	17
Chapter 3: How are colds transmitted?	28
Chapter 4: What makes someone more susceptible to a cold?	49
Chapter 5: Is the indoor environment an incubator for colds?	72
Section 2: The Discovery of the Cold Viruses	89
Chapter 6: The discovery of influenza	91
Chapter 7: The hunt for the 'cold virus'	116
Chapter 8: Robert Chanock joins the hunt	136
Chapter 9: Enter the coronaviruses	150
Chapter 10: Virus hunting in the molecular age	175
Chapter 11: COVID-19	195
Section 3: What can we do about colds?	215
Chapter 12: Over-the-counter medications	217
Chapter 13: Personal protection	241
Chapter 14: Air hygiene	259
Epilogue: We can do better than the Two-hat Remedy	278
Acknowledgements	285
Notes	287

Introduction

There's something in your throat. Somewhere around the back of your palate, between your tonsils. You swallow, you cough a little, maybe you try to spit it out but it's still there. It's not painful, but your heart sinks because you know what comes next. By tomorrow, your throat will be burning and over the next two or three days, that irritation will have found its way up into your nose.

You have a cold.

If you're lucky, you're in for nothing worse than a few days of sniffling, but you know there's a smorgasbord of miseries that the little irritation might turn into. Maybe it will reduce your voice to a croaking whisper forced through a swollen throat. Maybe it will cause you coughing fits that pound your diaphragm like a drumskin. Maybe it will drain the strength from your limbs and turn you into a boneless puddle on your sofa.

You're not looking forward to any of it but you're not particularly afraid. It's just a cold. You've had more colds than you can remember and while some have made you thoroughly miserable, you've always recovered within a couple of weeks.

That's how I thought of colds when I was studying how the lymphocyte cells of the immune system respond to life-changing and often life-ending infections like HIV and tuberculosis. Colds were annoyances to be sniffled through with the help of medications from the local pharmacy, not research subjects worth pursuing.

Back then, I was fortunate enough to be reasonably healthy.

My own lymphocytes adjusted my attitude to colds by trying to kill me. Some of them mutated into a type of cancer called a lymphoma and after the malignant cells survived several rounds of chemotherapy, the doctors decided the only option was to get rid of my lymphocytes altogether.

All of them.

I needed a procedure called a haematopoietic stem cell transplant. A high dose of chemotherapy cleared out all of my lymphocytes, along with several other types of immune cells, and then they were replaced with a donor's. It worked; there has been no sign of that cancer since. But it's left me with an immune system that doesn't work very well.

Colds are no longer a minor annoyance. They are my nemesis. They've sent me to A&E several times and even when I avoid the linoleum-floored waiting rooms, I'm often left ill and exhausted for weeks at a time.

Yet, compared to some people, I'm one of the lucky ones.

I've spoken to many people whom colds have seriously harmed, but of all the stories they've told me, it was Lucy's[*] story that epitomised how much damage they can do to someone's life.

In 2018, Lucy was a 37-year-old fitness instructor. She was as healthy as her job demanded but then she got the flu. Most people regard the 'flu' as just a cold that's bad enough to take time off work – but technically speaking, influenza is one of the many classes of cold virus and, like every cold virus, an influenza virus can do much worse than just a cold.

Every winter, the influenza virus cuts a swathe of coughing and sneezing across the northern hemisphere temperate zone, including Lucy's home city of York. When it infected Lucy, it did worse than make her cough and sneeze. Much worse.

It interfered with the part of her nervous system that regulates her heart. Five times that year, tachycardia sent her heartbeat racing so dangerously out of control that she had to be hospitalised. Her doctors helped her to manage her condition, but she'd never lead a spin class again. She tried working as a hotel receptionist but had to give it up because simply standing up would bring on the tachycardia.

Worse was to come.

The COVID-19 pandemic brought worse for everyone but Lucy's dysautonomia, to give her neurological condition its technical name, left her particularly vulnerable. She had no option but to shield,

---

[*] Surname withheld.

imposing lockdown restrictions on herself long after they ceased to be mandatory. She succeeded in remaining a 'novid', as the ever-dwindling holdouts from COVID-19 became known, as wave after wave of successively more infectious variants swept around the world. Lucy's luck ran out in August 2023, by which time the relatively mild Omicron variant had replaced the more devastating earlier variants.

Lucy's first experience of COVID-19 was mild only in the clinical sense that it didn't send her to hospital. It left her so ill that she had to sleep on her sofa for five days because she couldn't climb the stairs to her bedroom. Nevertheless, she made a full recovery. The first bout of COVID-19 is something of a baptism of fire because it carries the highest risk of serious complications[1] but, as Lucy would prove, the lower risk carried by subsequent infections is still a risk.

Her second bout, in May 2024, took her from waking up with a sore throat to being ambulanced to hospital within a few hours. Her heart problems 'did a one-eighty'[2], as she put it to me. Before, the tachycardia would hit her when she was standing up, which she could deal with by lying down. Now, her heart had started racing when she was lying down which was much more difficult to control. That second COVID-19 infection had inflamed her heart muscle, which added another layer of complication to the harm that influenza had already done to the nervous regulation of her heart. To add yet more complication, the inflammation spread beyond her heart to induce allergies she'd never had before.

By January 2025, when she told me what influenza and COVID-19 had done to her, she was describing a body launching a multifaceted attack on itself. By then, such complex conditions were becoming much better recognised as a common feature of 'long covid' than they had been when Lucy first experienced them as a long-term consequence of influenza, although effective treatment remains elusive. Now in her early forties, Lucy rarely gets through four weeks at a time without a hospital visit and she often needs a mobility scooter to leave the house[3].

Lucy's experience is a far cry from a cold in the usual sense of the word, although 'cold' is not a precisely defined term. In medical notes, colds are often described as 'coryza', which is a technical term for inflammation of the sinuses. The term doesn't say anything about what's causing the condition and it doesn't capture the sore throats, the headaches, the aching muscles and the plethora of aches and pains that come with that coryza.

In the absence of a satisfactory definition for something that anyone from a professor of medicine to a 12-year-old schoolchild is perfectly capable of recognising, I shall exercise the privilege of an author to define what the book is about. Within these covers, a cold is an illness caused by a virus that infects the upper respiratory tract and, in most cases, clears up within a matter of days or possibly weeks without requiring medical attention.

That definition is the reason that I include the influenza viruses and the SARS-CoV-2 virus that causes COVID-19 as cold viruses. They are more likely than any of the other cold viruses to cause the sort of serious illnesses that affected Lucy but that's a matter of degree rather than kind; some infections with every type of cold virus lead to some sort of serious illness. Most people get through most influenza or SARS-CoV-2 infections with an illness they'd call a cold and will only know which virus is causing it if they test for it.

Nevertheless, stating my definition in a room full of medical scientists would trigger at least half an hour of spirited debate. The clinicians would point out that every virus that causes something that might be called a cold can also cause a potentially fatal illness in some people and long-term complications, such as Lucy's, in others.

How many cases count as 'most cases'? The virologists would point out that nobody knows exactly how many viruses cause colds, but there are over two hundred that clearly fit that definition[4] and several others that sort of fit it and sort of don't, so how will I decide which ones to include and which to exclude? The bacteriologists would name a few bacteria that cause conditions that fit that description and demand to know why I'm excluding them. The molecular biologists, few of whom are well-versed in anatomy, might ask what I mean by the upper respiratory tract?

To most of those questions, my answer would be an embarrassed shrug and an acknowledgement that while we can debate whether certain borderline cases count as colds, we all mean the same thing when we say we have a cold. That's the condition I'm going to concentrate on.

That last question, about what the upper respiratory tract is, warrants a better answer than authorial privilege. In simple terms, it's the upper part of the respiratory tract, which is the system by which air gets in and out of the human body. Everything above the base of the neck is referred to as the upper respiratory tract and everything below is the lower respiratory tract. As the illustration on page xiii shows, the upper respiratory tract includes the throat and the sinuses, which are where cold viruses tend to infect. In most colds, what starts in the upper respiratory tract stays in the upper respiratory tract. Throats may throb and noses may bung but as long as the infection and its consequences stay between the throat and the paranasal sinuses at the front of the forehead, they're unlikely to be too serious.

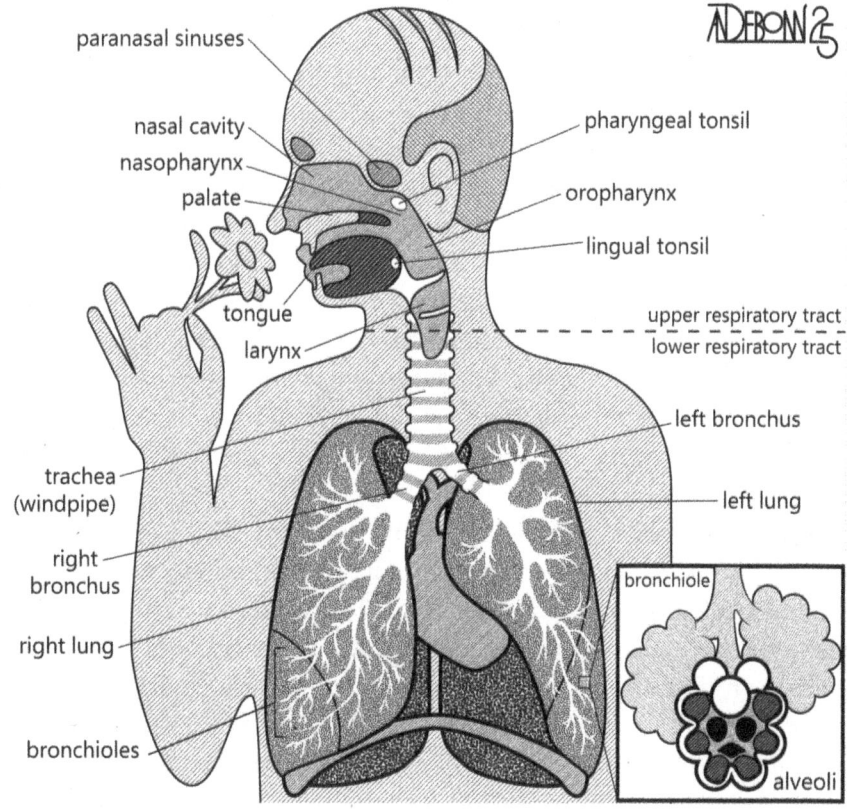

The anatomy of the respiratory tract. Air is drawn in through the nose and mouth, down the trachea and into the alveoli in the lungs, where oxygen is extracted into the bloodstream and carbon dioxide is passed back into the air to be breathed back out.

The really bad news starts when the mischief reaches other parts of the body. The virus may spread down the windpipe, follow the split into the two bronchi and infect its way along the branches of the 'respiratory tree', so called because the bronchioles split and split again like the branches and twigs of a tree. An inflamed throat is painful and unpleasant, but the throat is wide enough to allow space for a lot of swelling. The bronchioles are so narrow that the same level of inflammation and swelling can close them and block the passage of air to the lungs. If the viral infection reaches the alveoli, which are the structures in the lung that transfer oxygen from the air into the bloodstream, the inflammation and swelling spread with it and cause pneumonia.

However, the consequences of an infection can spread beyond the upper respiratory tract even if the virus itself doesn't make it below the neck. That's probably what happened to Lucy when she caught influenza in 2018. It's impossible to know exactly what the virus that infected her got up to, but it probably triggered inflammation that reached her heart and the nerves that regulate it. It could have caused her dysautonomia without infecting a single cell lower than her throat.

Such serious complications of cold viruses are exceptions, albeit much more frequent exceptions since the emergence of COVID-19. Nevertheless, most cases of influenza, COVID-19 or cold viruses of any other type don't cause anything worse than a cold. Colds are inconvenient, they're sometimes painful and yes, they come with a risk of causing something worse, but most of us see them as an inevitable fact of life.

They are a very expensive fact of life. In 2023, colds were estimated to cost the British economy around £44 billion, which works out to about £850 for every British worker*[6]. That could have paid that year's entire British defence budget† and the estimate only counts the sort of short-term illness for which the worst-case scenario is a few days off work. It wasn't even trying to capture the economic impact of long-term illness that turned a tax-paying fitness instructor like Lucy into

---

* In 2023, £44 billion was equivalent to $56 billion or €51 billion, and £850 was equivalent to $1,080 or €990.[5]
† See Chapter 5 for a more detailed description of how that £44 billion was calculated.

someone who needs regular support from hospital-based specialists to stay reasonably functional.

At the time of writing, the British evaluation is the most recent estimate and the only one calculated since SARS-CoV-2 has become a widespread cold virus, rendering the many pre-2020 estimates out of date. Those figures won't translate exactly to other countries because differences in population or employment structures will leave some national economies more likely to be impacted by colds than others. Nevertheless, a widespread economic burden that matches one of the world's ten most expensive militaries should give any national economist a sleepless night or two, even if they're not concerned about people becoming ill.

If you're not a national economist but someone like me, who understands money in layman's terms like grocery shopping, energy bills, rent and mortgages, you'll find it as difficult as I do to get your head around the astronomical figures by which national economies are measured. Yet all those billions are the sum of a cost that's shared among all of us as individuals.

If you're running a business, you pay for it in your employees' sick days and in the time they take off to care for an ailing child. If they drag themselves and their cold into work as 'presentees', you pay for it first in their slower rate of work and then when the employees they've infected fall ill a couple of days later.

If you're one of those staff, you'd better hope you have the sort of job that pays for your sick leave. If not, you'll have to choose between losing pay while you recover and being *that* person who drags themselves into work to share their cold with their colleagues.

That £44 billion is the money that cold viruses take out of all of our pockets.

Colds are a multibillion-pound burden on our civilisation, which is ironic because most of them are a product *of* our civilisation. The reason lies in the fact that colds clear up within a few days or, to get more technical, the fact that it only takes our immune system a few days to clear a cold virus out of our respiratory tract. After that, we will be immune to that particular virus for weeks or months or possibly years, depending on which virus happened to cause that particular cold.

Now and again, I've put some of these points to someone – not all of them at once, I'm not that much of a pub bore – to be told that sure, nobody likes having colds, but people have always had colds and always will. They've been part of being human for as long as there have been humans – so I'm often told.

It's not true. Cold viruses can only exist because we've spent the last few millennia building a society that enables them to exist.

Most cold viruses are only infectious for a few days at a time and they leave their hosts immune for a period that can last for weeks, for years or, in some cases, for the rest of the host's life. They can only persist in a large, interconnected society in which infected people pass close enough to so many other people that some are bound to be susceptible. The overwhelming majority of people alive today live in large and interconnected societies making it easy to forget that such societies are a relatively recent innovation in the hundreds of thousands of years in which our species has walked the earth.

Cold viruses are as much a product of the society we humans have built for ourselves as the publishing infrastructure and the electricity grid that brought this book to you.

We started building large societies between 10,000 and 15,000 years ago when our ancestors began the transformation from nomadic hunter-gatherers to denser and better populations supported by farming[7]. Those first farmers created an ecological niche for cold viruses to exploit and in the last few centuries, we humans have made that niche even more friendly for viruses. We've flocked to cities so populous that we can share our breath, and whatever viruses our breath carries, with hundreds of total strangers every day. We've connected those cities so efficiently that a cold virus exhaled in London's Heathrow airport today will be infecting people in Rome, Kinshasa, Karachi and a dozen other cities by the same time tomorrow and will be spreading into the rural areas around those cities by the day after that.

We've seen a century and a half of success built upon success in the field of infection control. Water is pumped around whole countries yet when it gushes out of taps, we drink it without sparing a moment's thought to waterborne infections like cholera and typhoid that decimated our ancestors. Vaccinations have all but eliminated many of the infections that filled all those child-sized graves that can be seen in every cemetery that goes back more than a century. Yet we remain as vulnerable to colds as our Victorian ancestors.

That's because the approaches that worked against so many other diseases won't work against colds. Providing clean water doesn't protect us against viruses transmitted through the air. Vaccines have reined in the worst depredations of influenza and COVID-19, but they give nowhere near the level of protection conferred by vaccines against diseases like measles and whooping cough.

Part of the reason is that a vaccine works against one microbe and, occasionally, one or two of its close relatives. It's simply not practical to come up with over 200 vaccines against over 200 cold viruses. Even if it were, vaccines against infections of the upper respiratory tract are always going to have their limitations.

That's because vaccines work by replicating the immune response with which the human body responds to a microbe. Some viruses, like those that cause measles and polio, induce immune responses that are very effective at protecting us against future infections. That gives vaccinologists an immune response to work with, enabling them to develop vaccines that trigger that immune response in children who have yet to be infected.

Immune responses to cold viruses are rarely so effective. There's a clue in the way that we are repeatedly infected with influenza viruses and SARS-CoV-2 despite the immunity that each infection induces. Both of those viruses evolve so fast that we rarely encounter the same virus twice. Someone who was infected with an influenza virus last winter is likely to have a reasonable level of immunity to that virus right now but when they encounter the descendants of that virus in a year or two, it will have accumulated enough mutations to slip past their immune system[8] and infect them again. SARS-CoV-2 evolves so much faster that she'll be lucky if her immunity lasts six months[9].

Vaccines can replicate those immune responses, but unless and until there's a bigger research breakthrough than is currently on the horizon, they're not going to improve on them.

If we're looking for something we can do about an illness caused by as many very different viruses as those that cause the common cold, we're not going to find it by repeating what's worked against other diseases. We can, however, learn a few lessons from past successes that may help us to innovate a solution. If we've prevented a myriad of different diseases by making food and water hygiene a societal priority, perhaps we can prevent cold viruses carried through the air by making air hygiene a priority.

If I stroll down my local high street, I'll pass rows of restaurants that display a hygiene rating in the window. Those hygiene ratings are there because anyone tempted by the menu is going to want to be confident they can tuck in without worrying about the food making them ill. However, few of us consider the hygiene of the air that we breathe or, to put it another way, how much of the air we breathe into our respiratory tracts has just come out of someone else's. Given how much of modern life is spent in indoor public spaces, that might be considered something of an oversight.

Air hygiene, to use the catch-all term for everything regarding the cleanliness of indoor air, is far from a new concern. It was first raised by Florence Nightingale[10] who was, among her many achievements, one of the most influential of the Victorian sanitary reformers. Nightingale had some success in persuading her contemporaries to design ventilation into new hospitals but since then, air hygiene has drifted in and out of vogue while our cities have spread and sprawled with buildings that, in all too many cases, might have been optimised to spread the airborne cold viruses.

The COVID-19 pandemic brought air hygiene back into vogue although, as I write this in 2025, the lessons of 2020 are already fading and, despite the ongoing risk of the sort of serious complications that Lucy has suffered, colds are once again seen as an inevitable and inconvenient fact of life.

That's unfortunate because we've never been better placed to do something about colds.

This book will cover the hundred years of research that got us to this point. The first section describes the research that taught us what a cold is and how we catch it, starting with the seminal experiments done in the 1920s and leading through the century of work that built on those experiments to bring us to the understanding we have in the 2020s.

The second section will describe the viruses that cause colds, focusing on the seven groups containing most of the known cold viruses. Getting to know such a diverse crowd has been a prolonged process. The first was discovered in 1933 when a couple of medical scientists isolated influenza viruses from their own sore throats. They had no idea of how many cold viruses there were to discover but they set the ball rolling. The rolling hasn't been particularly smooth; the process has often bumped against the limits of the virological techniques of the day. Nevertheless, the discoveries kept rolling in over the subsequent decades. A whole new class of cold viruses was discovered as recently as 2005[11] and there are certain to be more cold viruses yet to be found.

The third section will cover how a century of painstakingly – and often painfully – acquired knowledge shows us what we can do about colds. As individuals, we can alleviate our colds with over-the-counter medications and we can protect ourselves with handwashing and facemasks. As a society, there are ways to make our environment less conducive to catching colds.

Much of the history of cold research has happened out of sight. Cold researchers are among the backroom boys of medical science, rarely attracting the headlines and awards that attend success in more glamorous fields. Some cold researchers have been unsung heroes, quietly uncovering the secrets of one of humanity's most pernicious pests and sometimes developing techniques that drive breakthroughs in other fields. Others fall regrettably short of heroism.

Scientific institutions are as prone to such prejudices as any other part of the society in which they operate. If a society favours confidence born of ego over the constructive doubt that underpins meticulous research, then the institutional processes that decide which scientists get promoted and funded will be guided by that same prejudice. If a society harbours discrimination against those born into the wrong class, sex or ethnicity, that discrimination will pervade scientific institutions as much as the rest of society.

That human factor in its discovery demands some critical thinking. If a scientist's conclusion is built on solid evidence and sound reasoning, we can accept it irrespective of whatever prejudices that scientist might have held. If not, we must discard it even if we admire the scientist as an individual.

Noble and heroic scientists are often wrong. Critical thinking is the essential tool that sifts true knowledge from misleading factoids.

We can't always depend on heroes but in the science of colds, there's never any doubt about who the villains are: it's those pesky cold viruses.

Section 1

How colds work

Chapter 1

What is a cold?

Sometimes, scientific discoveries arise out of heroic quests for elusive truths. More often, they arise when a scientist notices a pattern and thinks to themselves: 'That's strange. I wonder what's happening there.' The discovery that the common cold is a viral disease fell into the latter category.

By the mid-1920s, Alphonse Dochez had spent 20 years studying pneumonia caused by bacteria infecting the lungs. He'd noticed that the first signs of pneumonia were often sore throats, coughing and runny noses. In short, pneumonia often began with a cold.

If nobody had noticed the pattern before, it may have been because a relatively trivial cold often slipped the mind of someone who had been prostrated with pneumonia. Before antibiotics were available, bacterial pneumonia would render a sufferer feverish and struggling to breathe for many weeks if they were lucky, or kill them in a few weeks if they weren't[12].

Dochez recalled asking one man if he'd had a cold and getting the reply, 'Oh no, I didn't have any cold or anything like that.'

He then asked the man's wife.

'Why, *did* he have a cold?' she replied. 'He's been coughing around the house for a couple of weeks.'[13]

Dochez noticed the pattern often enough that he mentioned it to the man who funded his research, Francis Garvan, president of a chemical company called Chemical Foundation Inc.

Corporate branding was more straightforward in the 1920s than it is now.

Garvan mentioned that his children often developed bacterial ear infections after a cold. Most colds were no more than an inconvenience for a week or two, but Dochez and Garvan agreed that a cold could herald a more serious illness.

'Why don't you study the common cold?' Garvan asked Dochez.

Dochez gave the answer that every medical researcher has given to such a question more times than they care to remember. 'We haven't the money.'

'Well, I'll give you the money,' said Garvan[14].

The hunt for the common cold was on.

Tentative hypotheses

Dochez's education was a product of a revolution in the scientific understanding of disease.

In the early 1870s, the idea that disease might be caused by microbes, meaning bacteria, viruses and any other living thing too small to be seen without a microscope, was a fringe theory. The few men who championed it were firmly outside the mainstream of medicine. By the early 1890s, the microbial theory *was* the mainstream of medicine and the most prominent of those outsiders, Louis Pasteur, Robert Koch and Joseph Lister, were celebrated as visionaries.

When Dochez started his medical degree in 1903, he joined the first generation of doctors and biologists to be taught the microbial theory from the day they began their studies. However, the microbial theory could only point a scientist in the right direction. It showed that disease could be treated by removing malevolent microbes from diseased bodies but to turn scientific theory into medical practice, a scientist needed to know which microbes caused which diseases, how they caused those diseases and how to remove them. Those were the blanks that Dochez's work on pneumonia was filling in.

At the time of that conversation with Garvan, Dochez was a professor of medicine at New York's Columbia University who, according to his colleagues, was so well dressed and 'extraordinarily handsome' that a passerby seeing him 'on Fifth Avenue one would never in the world suspect that one was looking at a distinguished biological scientist'[15]. Beneath his immaculate grooming lay an analytic mind far ahead of many of his contemporaries.

One of his assistants, who joined Dochez's team at around the time that Garvan offered to fund his cold research, later summed up his approach: 'He always maintained that if one had a thorough and selective knowledge of the literature one could find clues. These clues one sorted in one's head until some sort of tentative hypothesis emerged and then one devised a simple experiment to test it. If this yielded what he used to speak of as an "indication", then one repeated the experiment, and if the indication was still present one devised a larger and more formal series of experiments to establish a convincing theory.'[16]

It was a much more rigorous way to approach a knotty problem than that used by many of Dochez's peers which may explain why, at a time when most scientists believed colds were caused by bacteria, Dochez came to believe he was looking for a virus. The idea that bacteria caused colds was not illogical; Dochez himself had, after all, shown that pneumonia was usually caused by a bacterial infection. If bacteria caused a disease of the lungs at the bottom of the respiratory tract and that disease was often preceded by a cold in the nose and throat at the top, it made sense that the cold was the first sign of the bacteria that would go on to infect the lungs.

Dochez's belief that colds were viral was based on his knowledge of the literature, which encompassed a 1914 report by Walther Kruse at Leipzig University. One day, Kruse's assistant had come to work with a cold, so Kruse took the opportunity to collect some of the gunk dripping out of his nose. He passed the gunk through a filter fine enough to remove any bacteria and squirted the filtered liquid up the noses of his medical students. Several students developed a cold, which suggested that whatever caused it was small enough to pass through Kruse's filter and therefore smaller than a bacterium.

Early-twentieth-century medical researchers were aware of the existence of an 'ultra-microbe'[17] much smaller than a bacterium, which they called a 'filtrable virus'. 'Filtrable' because while they had no way to see it directly, they knew it could pass through filters fine enough to remove bacteria. 'Virus' is Latin for poison, which they already knew to be a misnomer. Unlike a poison, the ultra-microbe could replicate itself which was characteristic of a living entity.

Knowing that a virus was a filtrable infective agent led Kruse to conclude that a virus was the agent of the common cold[18] but he never followed up that single experiment. A couple of years later, Harvard University's George Foster performed a similar experiment on American soldiers[19] with similar results. Unfortunately, Foster found some bacteria in the filtered material which sent him on a wild-goose chase. He believed he'd found a bacterium that grew from some tiny seed small enough to pass through a filter. There was no such seed; the bacteria were in there because Foster's filters were not as fine as he believed.

Despite Foster's confusion, Dochez saw Foster's and Kruse's work as evidence that colds were caused by one of those malevolent ultra-microbes that could only be detected by the disease it caused. However, he couldn't ignore the prevailing view that a cold was a bacterial disease. Several scientists had, after all, shown that the nasal discharge from people sniffling with colds was teeming with bacteria. If there were a lot of bacteria up a diseased nose, they concluded, those bacteria must be causing the disease.

It qualified as a 'tentative hypothesis' but, as happened all too often at the time, several scientists prematurely concluded that they had proved the hypothesis and called it a theory. If bacteria were associated with disease and people with a disease of the nose had a lot of bacteria in their noses, they reasoned, the bacteria must be causing the disease. Their critical mistake was that they assumed that if someone showed symptoms of a disease in their nose and there were bacteria up that nose, the bacteria must be causing the symptoms. They didn't check that there were no bacteria up the noses of healthy people.

This is where Dochez showed his approach to be ahead of his peers. He did not mistake a hypothesis, which needs to be tested, for theory, which is constructed from evidence. Nor did he mistake assumption for evidence.

Instead, he tested the assumption that there was anything abnormal about having bacteria up one's nose. He persuaded his colleagues to be sampled every week over several months whether or not they had a cold. That's one way in which the common cold is easier to research than most other infections: it's so common that a researcher only has to follow a few people for a few months to see some of them catch a cold to study.

When Dochez's volunteers caught a cold, their noses and throats were indeed teeming with bacteria – but no more so than when they didn't have a cold[20]. If a bacterium was causing a cold, he would have expected to see some change in the 'nasal flora' associated with a cold. Samples from people with colds might have contained some bacterium that only appeared when someone had a cold or they might have had a markedly higher concentration of bacteria, but Dochez found no difference at all between the bacteria up a cold-afflicted nose and a healthy nose.

Of colds and chimpanzees

With the bacterial hypothesis put to rest, Dochez returned to his original idea that he was looking for a virus. He followed Kruse's and Foster's approach by filtering nasal discharge from volunteers suffering from colds and dripping it into volunteers who were willing to suffer a cold. His results were inconsistent; sometimes he could infect people and sometimes he couldn't. Knowing what we know today about the many different cold viruses and the different ways that different people respond to them, that's not particularly surprising.

Dochez thought he was looking for a single type of virus so he concluded that the inconsistencies lay with the people he was trying to infect. Human beings are awkward subjects for an experiment. It's much easier to experiment on animals that can be confined to a cage where a scientist can control what they eat and with whom they mix. Humans tend to object to that level of control.

However, viruses tend to be very picky about which animals they infect and Dochez couldn't find the right combination of virus and animal. He began to wonder if the animals he was working with were simply too different to humans to be infected with a human cold virus.

Dochez needed an animal as similar as possible to a human being, which got him thinking about chimpanzees. He spoke to keepers at New York's Bronx Zoo, who told him about the 'Monday cold'. The zoo's chimpanzees often developed runny noses on a Monday, which the keepers put down to their catching colds from visitors during busy weekends[21]. The keepers added that they often caught the Monday colds from their chimpanzees.

It sounded like human cold viruses could infect chimpanzees as readily as humans, so Dochez spent Garvan's money on setting up a chimpanzee laboratory. He found that if he isolated and filtered nasal gunk from someone with a cold, it would cause a cold in around half his chimpanzees which was, once he'd got the technique refined, similar to the proportion of human volunteers to whom he could give a cold.

Dochez published his final results in 1930[22], six years after starting that first experiment that would show that colds are caused not by bacteria but by viruses. His discovery opened the door to everything that has since been discovered about colds.

It was one thing to show that a cold was caused by a virus. It was another thing altogether to understand *how* a virus causes a cold.

The common cold constitution

The human body is home to hordes of microbes. So many that we have at least as many bacterial cells as human cells although, because bacterial cells are so much smaller than human cells, they only make up two or three grams in every kilo[23]. It's not only bacteria that call us home; most of us are carrying around a few viruses as well[24] and we get along perfectly well with most of our tiny passengers.

When a microbe causes a disease, the disease arises from the interaction between that microbe and the human body's response to it.

While Dochez was working on the microbial side of that interaction, William Gafafer and James Doull at Baltimore's Johns Hopkins University were looking into the human side. Between 1928 and 1932, they recorded how many colds their students suffered. Around half had two or three colds every year. However, a quarter had at least four with one unfortunate student recording nine colds in a single year. Another quarter was luckier, with only one or none at all. The differences they found were consistent; students who got a lot of colds in one year usually got a lot of colds in another while some students breezed through year after year untouched by all the colds going round[25].

Among any group of people who work or study together, it soon becomes obvious that some catch most of the colds while others remain blissfully – and, to the oft-afflicted, infuriatingly – untroubled by colds. Gafafer and Doull's students were all studying in the same lecture theatres and libraries so whenever a cold virus got loose among them, they were all exposed to it. They concluded that some students caught more colds than others 'because of possessing differing powers of resistance', but a 'power of resistance' was a nebulous concept. It didn't explain *how* their students were resisting colds and why some resisted colds so much better than others.

Gafafer and Doull's search for a deeper explanation appears to have been guided by the paradigm of eugenics, which pervaded American scientific and political circles in the 1920s and 1930s. Eugenicists believed that a human is born with an immutable set of characteristics inherited from their ancestors, making some people inherently superior to others. Being resistant to colds looked like a superior trait so they looked for other inherited characteristics that might go with it. They didn't find any, although they did establish that cold resistance had nothing to do with their students' eye colour[26] or whether they were Jewish[27]. Eugenic explanations never stand much scrutiny.

However, their initial observation was sound and has been confirmed by many studies since: some people are simply more likely to develop a cold than others. In 1947, a Boston University team introduced the idea of the 'common cold constitution'[28] which avoided the eugenic assumptions that power or resistance had anything to do with it. Its very vagueness made 'common cold constitution' a more apposite term than 'power of resistance' because it acknowledged that nobody knew what made some people more cold-prone than others.

To get a handle on the mechanisms underlying it, someone needed to work out what happens when a cold virus gets up a nose.

A piece of bad news wrapped up in protein

One of the more succinct descriptions of a virus is that it's a 'piece of bad news wrapped up in protein'[29]. The bad news is a fragment of genetic material which, like the genetic material of plants, bacteria, fungi, beetles, herrings, tapirs, you, me and every other self-contained living thing, is encoded into a type of biochemical called nucleic acid. The nucleic acid that carries the genes of all those living things is DNA, short for deoxyribonucleic acid. Some viruses also use DNA while others use a similar molecule called ribonucleic acid, or RNA.

A virus perpetuates itself by inserting its DNA or RNA into a living cell, where it hijacks the cell's molecular machinery to replicate itself. That's why viruses are not self-contained living things; they can only continue to exist by making use of other living things. It's also what makes a virus bad news. We need our cells working for us, not for the rogue genetic material that is a virus.

Outside a living cell, a virus's DNA or RNA is completely inert. It can't do anything by itself, including getting into another living cell. Yet a virus could not be viral if, having replicated inside a cell, those replicas were stuck inside that cell. They need to get out and find another cell to infect, whether that cell is inside the same body or in another. A virus emerging from one of my cells may infect another one of my cells or it may infect one of yours.

That's where the protein wrapping comes in.

The capsid, to give the wrapping its technical term, protects the genetic material on its travels but it's also how the genetic material ends those travels. It attaches itself to the outside of a new cell and inserts the genetic material into it so the virus can start another replication cycle – but not just any cell.

The protein molecules of a virus's capsid are honed to attach to a single type of protein. Different cell types express different protein molecules on their surface, so a particular virus can only infect cells expressing that target protein. Moreover, a virus has to bump into a cell before it can attach to it, and some cells are more available to bump into than others. Hence there are mercifully few viruses that infect our brains because our brains are behind multiple barriers.

On the other hand, the cells lining the nose and throat are as exposed to the outside world as human cells can get. A typical adult inhales and exhales around half a litre of air with every breath, adding up to between 10,000 and 20,000 litres of air every day[30]. For comparison, a standard bathtub holds around 100 litres. A couple of hundred bathtubs is a huge amount of air passing through the respiratory tract in a day, bringing with it all the dust, fungal spores, pollen and whatever else is floating around in it. If you rarely think about your breathing – which you probably didn't until you started reading this – it's because your respiratory tract is very good at cleaning the air you're breathing in.

The respiratory tract is a microbial motorway into the human body for any virus that can avoid the mechanisms of cleaning. Our need to breathe air sets the stage for the battle between the cold virus and our body's defences and out of that battle arises the common cold.

The battle up the nose

The lower respiratory tract is protected by a conveyor belt of mucus that travels up the respiratory tract to the mouth. It's a lethal trap for a virus because it's laced with antiviral chemicals and as soon as mucus reaches the mouth, it gets swallowed down to the acid of the stomach along with anything caught in it[31].

That constant flow of mucus is why there are over 200 cold viruses that infect the top of the respiratory tract but very few that infect the lungs at the bottom. There are simply more opportunities for viruses to bump into cells at the top.

The upper respiratory tract may be more accessible than the lungs, but it is not undefended. One defensive mechanism is the nasal cycle: the tissue lining each nostril periodically distends as liquid is pushed out of the blood vessels and into the nostril. The distention waxes and wanes over a cycle of a few hours, which can be measured in the changing airflow through a nostril. One effect of the cycle is that at any given time, there is more air flowing through one nostril than the other although both nostrils cycle independently. The difference occurs simply because it's unlikely for both nostrils to be at the same stage of the cycle at the same time.

The liquid forced out of the blood vessels appears to be one of the upper respiratory tract's cleaning mechanisms, flushing anything hanging on to the inside of the nostril into the mucus[32]. As the figure on page 12 shows, it gets rid of fungal spores, pollen grains or indeed viruses that have yet to enter a cell.

When a virus does get into a cell, however, the infection triggers a much greater distention as the nose tries to flush it out[33]. The problem is that the virus is not on the lining of the nostril but already inside a cell, from where no amount of flushing will dislodge it.

The figure also shows that when flushing doesn't work, the immune system activates its first line of defence. It floods the upper respiratory tract with granulocytes[34], a type of white blood cell that's very effective against incipient fungal or bacterial infections but, like nasal flushing, cannot root a virus out of a cell that it has already infected. The distended blood vessels and granulocytes are key elements of the inflammatory response. When we get a splinter in a finger, it's the inflammatory response that makes the finger swell up and feel warm to the touch. Inflammation protects us against microbes driven into a finger but it's deeply unpleasant when it happens inside the nose and throat, especially when it's a reaction to an infection that the components of the inflammatory response are impotent against.

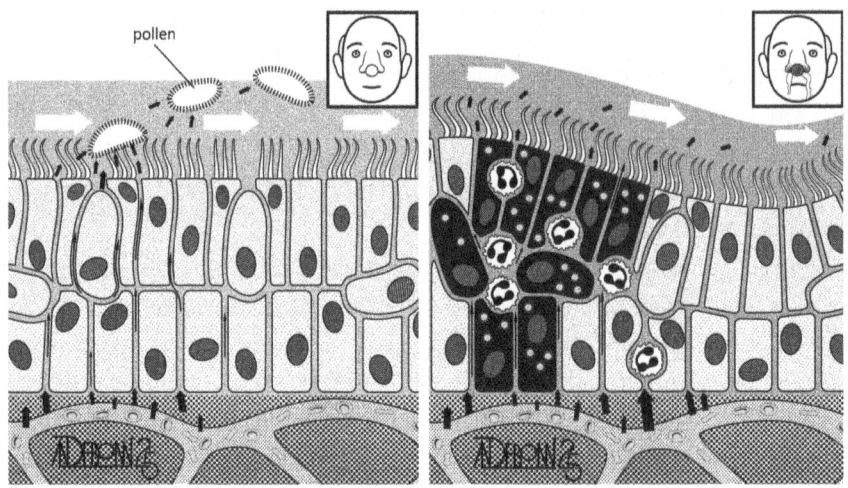

The left-hand panel shows how the nasal epithelium responds to pollen grains or other irritants on its surface, by increasing the pressure in the underlying capillaries to flush them into the mucus layer. The right-hand panel shows how the same response is overstimulated when a cold virus infects the cells of the epithelium. The flushing is ineffective because the virus is inside the cells instead of on the surface and when the stimulus persists, the granulocytes infiltrate the epithelium as part of the inflammatory response that we recognise as a cold.

Inflammation triggers coughing and sneezing, which are yet more ways of getting rid of irritants outside cells that don't do anything about viruses replicating inside them. Coughing and sneezing fling newly replicated viruses into the air from where someone else can inhale them but they do nothing about the infection creating those new viruses.

Inflammation is driven by various cells frantically signalling to each other that there is something to get inflamed about, and much of that signalling is done by secreting soluble protein molecules that other cells pick up. Sometimes there are so many of those molecules floating around that they get into the bloodstream and trigger inflammation that's nowhere near any virus[35]. If you've ever felt your whole body ache when you get a cold, that's why.

That's how a full-blown cold plays out but as Gafafer and Doull pointed out, not every infection causes a full-blown cold. There are some broad differences. A young child is much more likely to develop a fever than an adult. Men and women tend to experience colds slightly differently, probably because women tend to have a stronger cough reflex[36], leaving women more afflicted by coughing than men, who tend to have runnier noses[37].

Those differences should be interpreted with a caveat: they're based on averages drawn from observations of many colds in many men and women. Now and again, most men can expect to end up coughing like a pneumatic drill, just as most women will occasionally experience a nose running like the Victoria Falls.

The exceptions are the people blessed with extraordinarily potent powers of resistance, such as those students whom Gafafer and Doull observed getting through a full year without a single cold. Some five decades after they made that observation and four decades after the concept of the common cold constitution took hold, Jack Gwaltney's team at Charlottesville's University of Virginia started digging into the mechanisms behind the common cold constitution.

Gwaltney first got interested in colds in 1960, when he was an army medic responsible for an infirmary full of recruits left struggling to breathe by a type of cold virus called the adenovirus[38]. It was a stark demonstration that a cold can be a lot more serious than a sniffle, and when he left the army for the University of Virginia in 1962, he studied the common cold until he retired in 2000[39].

By the 1980s, scientists building on Dochez's work had established that there was not one common cold virus but many, which will be covered in detail in the second section of this book. The days of filtering someone's nasal secretions and basing an experiment around whatever happened to be in there lay in the past. When Gwaltney asked for volunteers to be infected with a cold, he knew exactly what he was infecting them with and could check they were not already immune to it.

Some of Gwaltney's volunteers developed colds, which his team confirmed by washing out their noses every day and checking the viruses in those washes were replicas of the viruses they had infected them with. However, they also found viruses up the noses of volunteers who showed no sign of having a cold. Their cold virus had infected them and was replicating merrily, but those volunteers were not producing the inflammatory response that was making their fellow volunteers ill.

Gwaltney's team found another difference between the two groups: the volunteers who developed colds were secreting a signalling molecule called bradykinin while the virus-infected but happily cold-free volunteers were not[40]. Bradykinin induces the blood vessels to expand and become more permeable, ineffectively flushing liquid from the blood into the nostrils and cranking up the flushing to painful levels when the irritant remains stubbornly unflushed. No bradykinin, it appeared, meant no inflammation and no cold despite their being thoroughly infected.

Gwaltney and other researchers have since built on the bradykinin studies, showing a whole range of signalling molecules associated with inflammation associated with colds and at least some of them are, like bradykinin, absent in people who are infected without becoming ill[41].

Gwaltney's results show the difference between *catching* a cold and *developing* a cold. Whether or not one has a common cold constitution depends not on whether one can avoid receiving the bad news of a cold virus but on how one responds to that bad news. Someone whose immune system responds with inflammation will suffer a cold while someone whose immune system leaves it alone will not.

There's almost certainly more to the common cold constitution than has been discovered so far and it's still unknown whether bradykinin or any of the other molecules explain the sex differences in susceptibility: boys tend to be more susceptible than girls but the balance flips at puberty, with women being more susceptible than men[42].

As a frequent cold sufferer myself, I'll admit to some frustration at the knowledge that they are effectively an exercise in self-harm. I find myself imagining the cold virus cackling to itself at the devastation it stimulates me to wreak upon myself while it remains safely hidden.

I comfort myself with the thought that it will be laughing on the other side of its face when my lymphocytes arrive.

Unleash the lymphocytes

Inflammation is part of the immune system's first line of defence, which swings into action whenever the body detects something that shouldn't be there. It's a mechanism of unceasing vigilance that keeps us safe from countless microbes, usually without us even noticing what it's doing.

If a microbe causes us enough trouble that we notice it, it's because it has a way around that first line of defence. The lymphocytes form a second line of defence that takes a few days to swing into action but when it does, it's the end of the road for a cold virus.

While a cold virus is reprogramming a human cell into a virus replication factory, the cell is not entirely asleep at the switch. It's spiriting away some of the viral protein molecules that the virus coerces it into manufacturing and displaying them on its surface. Displaying those proteins not only warns the immune system that the cell has been subverted but also provides a sample of the virus subverting it.

The immune system's second line of defence is an array of different types of white blood cells called lymphocytes. Each lymphocyte is primed to respond to a particular protein, and we have so many different lymphocytes that when a cell signals that it's been infected, it's a near certainty that we'll have lymphocytes that can respond to at least some of the viral proteins being displayed – but not enough lymphocytes to take on the virus immediately.

We can't carry around enough lymphocytes to instantly join battle against every microbe we encounter; we'd need to sustain so many lymphocytes that we'd have no metabolism left for anything else. The second line of defence needs to be built up, which the lymphocytes do by dividing, dividing and dividing again. After a few days, the few lymphocytes that recognised the invader give rise to an army dedicated to rooting it out.

Like any well-organised army, the lymphocyte battalions are arranged into different specialisations. When dealing with a cold virus, the first into the fray are usually the killer T-lymphocytes which destroy infected cells in which the virus is replicating. They're backed up by B-lymphocytes, which secrete soluble molecules called antibodies into the bodily fluids. Like the B-lymphocyte secreting it, an antibody recognises one particular protein. Antibodies can't get inside cells to attack a virus while it's replicating, but once the virus is outside the cell, those antibodies bind its capsid before it can get into another one.

The B-lymphocytes and killer T-lymphocytes mount a two-pronged attack, seeking out the invading virus both while it's replicating in a cell and while it's looking for a cell to replicate in. The combination usually stamps out the viral infection within a few days.

Once the infection is gone, the activated lymphocytes do not return to the quiescent state in which the infection found them. Lymphocytes retain a 'memory' of any infection that has activated them, which manifests differently in different types of lymphocytes. For the killer T-lymphocytes, memory enables them to divide much faster if the same virus infects for a second time. For B-lymphocytes, it means they continue to secrete antibodies that stop the same virus from infecting again.

For a doctor or a medical researcher, the presence of antibodies to a particular virus reveals that an individual has been infected by that virus in the past. That's how Jack Gwaltney was able to confirm his volunteers had not previously encountered the cold virus with which he infected them.

Immune memory – or lack thereof – is why children suffer more colds than adults[43]. The virus causing every cold leaves the immune system with a memory of it and as we get older, we accumulate immune memories that protect us from more and more cold viruses. Immune memory is not a panacea; as every cold virus must find a way around the first line of defence, some have found a way around the second. Section 2 will cover how different cold viruses each have their own bag of tricks to deceive or evade immune memory.

The lymphocyte response does not appear to have anything to do with the common cold constitution, which depends on the factors in play before the lymphocytes are mobilised. The killer T-cells of someone with an inflammatory response ferocious enough to confine them to bed purge the cold virus as effectively as the killer T-cells of someone blissfully unaware of being infected.

Today, we still don't know every detail of how a cold works, but we do know that a cold arises from the interaction between a cold virus and the human body it infects. However, it can only arise because a cold virus can get out of one infected human and infect another despite being completely inert while it is in transit.

A cold virus's continued existence depends on infected people passing close enough to susceptible people to infect them. For most cold viruses, that infection must happen within the few days before the lymphocytes shut down the virus's replication. The longer a cold virus has been in circulation, the more people will have acquired immunity to it and the more people an infected person will have to pass close to before one of them is susceptible enough to be infected. That means that colds can only persist in societies where every individual regularly passes close to a lot of other individuals, which in turn means that colds are not only a product of the interaction between a virus and a human body. They are also a product of our society.

Chapter 2

When did the first person catch the first cold?

Imagine a boy beside a river. He's between ten and twelve years old and, under the layers of fur protecting him from the freezing Arctic wind, his lean body is edging toward adulthood. So is his mouth, where his adult teeth are pushing his milk teeth into formations as jagged as the ice floes drifting down the river.

We don't know if he attached any particular significance to losing his milk teeth. Maybe his culture had its equivalent of the tooth fairy or maybe he discarded his teeth without a second thought. We do know that his teeth ended up buried in the silt deposited by what is today the Yana river in northern Siberia, where the climate is so cold that those teeth froze into the permafrost that reaches several metres below the surface of the Eurasian tundra.

That scene played out twice in the same place, which was probably a seasonal camp[44] used by generations of mostly nomadic hunter-gatherers. The permafrost froze two different milk teeth from two different boys who may have been playmates or may have been separated by more than a hundred years. That is all that's known about the two boys who left us one tooth apiece, whom palaeontologists call Yana1 and Yana2 for want of any other name.

In 1993, according to a calendar that would have mystified Yana1 and Yana2, something beside the Yana caught the eye of a geologist called Mikhail Dashtzeren. He looked a little closer and found a spear-shaft made from the horn of a woolly rhinoceros, a species that has now been extinct for at least 10,000 years but would have been a common sight for the Yana boys.

A team from the Russian Academy of Sciences reasoned that where there was one ancient human artefact, there might be others. They were right; their excavations unearthed a treasure trove of human prehistory.

Tools knapped from slate and quartz were buried along with the bones of brown bears, mammoths and more woolly rhinos that showed the marks of having been butchered with those tools[45] – and one milk tooth each from Yana1 and Yana2.

They had been frozen for between 30,000 and 35,000 years.

Ancient adenoviruses

The Yana boys' teeth were so pristine that the palaeontologists wondered if the permafrost might have preserved more than just the hard structure. Perhaps it had stabilised the boys' DNA and, just possibly, DNA from some of the viruses they happened to be carrying. The Russian Academy did what archaeologists and palaeontologists often do when they suspect an artefact may carry interesting DNA: they called Martin Sikora at the University of Copenhagen.

Extracting DNA from a 30,000-year-old tooth is not an easy task. If there's any DNA at all, it will be in such tiny quantities that one misstep can lose the irreplaceable lot. Sikora's laboratory specialises in extracting minuscule amounts of DNA from excavated bones, teeth and anything else that might carry it. He and his colleagues struck lucky with those milk teeth. Not only did they find Yana1's and Yana2's genetic codes, which is how we know they were boys[46], but they also found the DNA of five different viruses that still circulate today.

Four belonged to a group called the herpesviruses that rarely cause illness but establish lifelong infections. The fifth was a group C adenovirus, which typically causes a cold[47]. It's the oldest example of a cold virus ever found, and it shows that people have been catching colds for at least 30,000 years.

It's the first marker on the trail of the first cold viruses, although it's not the head of the trail. If the same virus turned up in two teeth from two different individuals, it must have been a common infection by the Yana boys' time. As with every significant discovery, the adenovirus DNA in those teeth raises at least as many questions as it answers.

One such question is whether Yana1 and Yana2 caught as many colds as we do. The answer is almost certainly not. The Yana boys lived at a time when all humans lived in hunter-gatherer societies that were too thinly spread to sustain most of the cold viruses in circulation today. For the Yana boys, a cold would have been a much more unusual experience than it is for a modern child.

Most human viruses are descended from animal viruses[48]. When you get a cold, you're playing temporary host to a virus whose ancestry goes back to some incautious individual who took a breath a little too close to an infected animal. Now and again, one of our hunter-gatherer ancestors must have picked up such a virus from an animal they hunted and butchered. However, hunter-gatherer societies would have been dead ends for most of the viruses that infected them.

Most cold viruses only have a few days to find a new host before the immune system stamps them out, leaving that individual immune to the virus for months, years or possibly for the rest of their life, depending on which cold virus infected them. Cold viruses can only persist in large, interconnected societies that enable them to pass from one non-immune person to another every few days. Among hunter-gatherer bands, a virus could only persist if it was infectious for a long enough period to cover the intermittent contacts between those bands.

The adenovirus infecting the Yana boys has a characteristic that makes it very unusual among the cold viruses. It's able to enter a state called latency, in which it lurks in the body for several months after the coughs and sniffles have subsided[49]. A latent adenovirus doesn't draw the attention of the immune system but it can still infect someone else.

Being infectious for months rather than days enabled the adenoviruses to persist in a thinly spread hunter-gatherer society. That adenovirus could bide its time until one of those intermittent contacts between groups offered it new hosts. The influenza viruses, coronaviruses, rhinoviruses and the various other cold viruses we'll meet in Section 2 cannot wait around for that long, making it likely that the adenoviruses were the only cold viruses that our hunter-gatherer ancestors had to contend with.

Nobody has calculated how large a human population is needed to sustain any of the cold viruses. However, minimum host population sizes have been calculated for other infections with the same need to rapidly and repeatedly pass from one non-immune person to another. Measles needs at least half a million people and whooping cough requires at least 200,000[50]. Both of those transmit in the same way as the cold viruses, so we can infer that the non-adenovirus cold viruses also need at least hundreds of thousands of people intermingling with one another. In the Yana boys' time, before humans congregated in those numbers, the adenoviruses were the only cold viruses that remained infective for long enough to persist.

It would be fascinating to know how long our ancestors had been passing adenoviruses around by the time the Yana boys caught them, but we have no way to find that out. While 30–35,000-year-old humans might seem ancient to us, they were relative latecomers to the human story.

Our ancestors' immune systems have been protecting us from potentially dangerous microbes, with varying degrees of success, since well before our furry ancestors were burrowing beneath the toenails of browsing dinosaurs tens or hundreds of millions of years ago. Since then, our ancestors got bigger, stood up on two legs and became something we might recognise as human. It's unclear when exactly that happened, but in Jebel Irhoud, Morocco, palaeontologists unearthed the remains of humans from about 300,000 years ago[51] who would not attract a second glance if they donned a suit and tie and walked into a Starbucks.

The first human adenovirus could have emerged at any point in the quarter-million years between the Jebel Irhoud humans and Yana1 and Yana2, or it could be even older; the first of our ancestors that it infected may be so far back that they predate our very humanity. We have no way to know.

Following the cold virus trail forward in time from the Yana boys leads us to another question: what changed to allow all those other cold viruses to join the adenoviruses in causing colds?

That's a question we can answer because long after the Yana boys lived, there was indeed a dramatic change that enabled cold viruses to sustain themselves – but it wasn't the viruses that changed.

It was us.

The Neolithic Revolution

The greatest upheaval in human history happened when our ancestors took up farming.

The first signs of what scholars of human prehistory call the 'Neolithic Revolution' appeared in the Middle East between 11,000 and 15,000 years ago[52]. In the subsequent millennia, people in several different parts of the world independently started their own experiments with agriculture. After hundreds of thousands of years of hunting and foraging, humans in the Indus Valley, Central America, the South American Andes, West Africa and possibly several other places started farming crops and livestock.

Farming didn't replace hunting and gathering overnight. It probably took several generations of experimenting with herding and planting before agriculture yielded more food than hunting and gathering. It took even more time for farming techniques to spread and develop.

But spread they did, not because agriculturalists were better fed than hunter-gatherers – early agriculturalists tended to be shorter than those of hunter-gatherers, suggesting they weren't as well nourished – but because they were far more fecund[53]. Agriculture concentrated food production, which allowed humans to concentrate their populations because they didn't need to cover as much ground as hunter-gatherers to get enough food. It enabled farmers to form larger, denser and more complex societies, but it also enabled higher fertility rates that drove a relentless need for more land to farm.

Hunter-gatherers who didn't take up farming were displaced. Their sparse populations simply could not resist the denser and better-organised agriculturalists' need for more land.

The Neolithic Revolution was a dramatic shift in how humans interacted with their social and physical environment but, from a microbial perspective, human beings were part of the environment that was changing. Humans now lived in much larger groups than they ever had before and they were even closer to the animals they were farming.

As societies expanded to encompass thousands of people, then tens of thousands and then hundreds of thousands, they ceased to be dead ends for viruses that were only infectious for a few days. If an animal virus infected a human, it now had a much better chance of finding an unbroken chain of susceptible people that could sustain it.

Moreover, it became far more likely that someone would be infected by an animal virus in the first place. Farmers spend far more time close to their livestock than hunters spend close to their quarries, giving any viruses those animals carry far more opportunity to infect them.

After hundreds of thousands of years of hunting and gathering, the Neolithic Revolution paved the way for diseases that could never have existed before[54].

The *Hippocratic Corpus*

If the non-adenovirus cold viruses could not have emerged before the Neolithic Revolution, it raises the question of whether we can establish when they did emerge more precisely than some time in the last 15 millennia. The answer to that question is a resounding 'not really'. We know our Neolithic ancestors through a small number of skeletons that span thousands of years, but cold viruses infect soft tissues that decay after a few years. They leave no mark on those skeletons for a modern palaeontologist to be able to say, 'aha, she's had a cold'[55].

After the Yana boys, the trail of the common cold goes cold for some 300 centuries. It only warms up when our ancestors started leaving more direct clues to their ailments than their bones and teeth: by writing them down. If the common cold annoyed those early scribes as much as it annoys us, surely we'd expect it to be among the ailments they wrote about.

The logical place to pick up the trail is in the earliest surviving medical text: the *Hippocratic Corpus*, written by the Greek physician and philosopher Hippocrates and his associates around 400 BCE.

Before plunging into the *Corpus*, a few words of caution are in order. It's easy to mislead oneself when trying to interpret past medical texts in the context of modern understanding of disease. The meaning of words often changes over time so when we encounter a medical term in an ancient text, we can't know whether what we understand by the term is what the author meant when they wrote it. Moreover, someone like me, who cannot read ancient Greek, depends on a translator's interpretation of the medical terms, adding yet more distance between me and the author's original meaning.

A further problem is that we can't assume the diseases described in historical texts are the same ones we recognise today. For example, the *Corpus* does not mention anything recognisable as smallpox or measles[56], both of which probably emerged several centuries later. On the other hand, Hippocrates and his contemporaries may have assumed their readers would be familiar with diseases caused by microbes that have since evolved beyond recognition or gone extinct, leaving us trying to interpret their descriptions with no idea of what they were describing.

All that uncertainty gives rise to an unfortunate certainty: if one were to dive into the *Hippocratic Corpus* in search of evidence that the authors were familiar with colds, one would find it whether or not the authors were writing about what we would refer to as a cold. There are bound to be descriptions of sore throats and stuffed noses because many different illnesses involve sore throats and stuffed noses.

With all that in mind, a description written between 410 and 400 BCE of a condition 'which we all experience' that involves 'movement of flux through the nostrils ... leading to hoarseness and sore throat'[57] but clears up in an unspecified but apparently short time, it's very tempting to read it as evidence that every ancient Greek had experienced a cold. Applying due caution, however, we must note that the Hippocratic description describes the symptoms occurring in the opposite order to that of most modern colds, which usually start with hoarseness and sore throat and then progress to nasal congestion[58]. The description goes on to talk about ulceration of the eyelid and 'burning heat and extreme inflammation', both of which sound much nastier than what we'd usually think of as a cold. The author *might* not have meant it in the dire terms with which we read it and *might* have meant the sort of sore eyes and slightly elevated temperature that we'd associate with a cold, or he *might* have been describing something altogether different from anything we'd recognise.

With all those caveats, the most we can say is that two and a half millennia ago, Hippocrates and his associates *might* have seen patients with what we'd recognise as a cold.

From County Kerry to *Wuthering Heights*

After Hippocrates's time, descriptions of ailments that might be colds turn up from time to time but, given the difficulties of interpreting historical texts, it's never clear whether a description of familiar-sounding symptoms is referring to what we'd recognise as a cold today.

One of the clearest marks was left in 1580 when an English army was, in what would become a longstanding tradition of English and subsequently British armies, waging war in Ireland. The English advance through County Kerry came to an abrupt halt when, according to one chronicler, 'suddenlie such a sicknes came among the soldiers' that spread through the army and left men lying 'as dead stockes, looking still when they should die'. After three days of prostration, most of the army recovered and 'such was the good will of God that few died'[59].

It reads remarkably like the influenza pandemics that tore around the world in 1918 and, in a less lethal but still dangerous form, in 1957 and 1968. Pandemic influenza viruses spread fast and make their victims so ill so quickly that in 1918, one British army medical officer described men joining a route march in the morning, collapsing halfway through it and dying by the end of the day[60]. The sixteenth-century description of an illness reducing scores of healthy men to a state of collapse 'at one instant'[61] sounds very similar to descriptions of influenza-stricken army barracks written about 1918*.

The way those Elizabethan soldiers either recovered relatively quickly or died is also characteristic of untreated pandemic influenza though, given what counted as medicine in sixteenth-century Europe, they were far more likely to recover if the physicians left them alone.

Another piece of evidence that the 'sicknes' was a pandemic influenza was that a similar illness was recorded from other parts of the world at the same time[62]. Whether or not the illness was indeed influenza, it does appear to have been a pandemic.

The chronicler who described that outbreak in Kerry went on to describe the disease spreading to England, where it was called the 'gentle correction'[63]. The name illustrates the difficulties of understanding disease in historical texts; without a surviving document to identify the gentle correction as a disease, a text describing the gentle correction arriving in England from Ireland would leave a modern historian perplexed.

The gentle correction is a significant marker on the common cold trail because the seasonal influenza viruses that sweep across the northern hemisphere every winter include direct descendants of the viruses that caused the pandemics of the past. Those relatively mild seasonal influenzas, which are effectively colds, were not reported until the late nineteenth century[64] although we can't know whether that's because there had never been seasonal influenza before that or simply because nobody had noticed it as distinct from other illnesses.

Irrespective of whether the gentle correction was a pandemic influenza or whether it became a seasonal influenza, it sounds very much

---

* The 1918 influenza pandemic and its relationship with the colds caused by seasonal influenza will be covered in more detail in Chapter 6.

like a highly infectious but short-lived infection that's at least similar to a cold virus. If the gentle correction became a pandemic, it was because human societies were sufficiently large and interconnected to sustain such a virus. It follows that they must have been sufficiently large and interconnected to sustain the non-adenovirus cold viruses.

The early modern era sees more descriptions of illnesses that are recognisably colds. In 1694, one Dr Molineux of Dublin sent a 'historical account of the late general coughs and colds'[65] to the Royal Society of London, describing various 'rheums' that afflicted his city every winter. Many of the illnesses he described read as much worse than a cold, with serious fevers, strong aversions to light and sound and death rates of around one per thousand. Yet Molineux also describes how, in December 1693, he couldn't enjoy going to church because 'the noise of coughing in every congregation was very troublesome, and never out of ones ears'[66].

Now *that* sounds like a cold as we know it.

Molineux described an illness affecting a lot of people at the same time, for which the most obvious symptom was coughing. Unlike the more serious afflictions that Molineux described, that cough wasn't debilitating enough to keep its sufferers from going to church. It seems safe to say that by the end of the seventeenth century there were cold viruses in circulation, although we don't know if they were as commonplace as they are now.

By the mid-nineteenth century, colds appear to have been a generally accepted fact of life because that's how they were described in the literature of the time. To take one of many examples, colds are constantly afflicting the characters of *Wuthering Heights*, Emily Brontë's gothic tale of star-crossed lovers who really should have taken the celestial hint. As well as an achingly poignant description of romance descending into domestic abuse, *Wuthering Heights* tells us that when it was published in 1847, colds had become as ubiquitous as they are today.

Brontë used a cold as a key plot device when Mr Lockwood, the narrator, is prostrated with a cold and housekeeper Nelly Dean tells him the story of Heathcliff and Cathy's childhood to keep him entertained.

It is striking that Brontë rendered Lockwood completely incapacitated by his cold. It sounds much worse than today's colds which, as debilitating as they sometimes are, we can usually drag around with us if we must. Lockwood suffered the fevers and weakness that appear in the *Hippocratic Corpus* and Molineux's versions of the cold that left people far too ill to be annoying in church.

It's difficult to interpret the fact that historical descriptions of colds so often sound much more serious than most colds we have today. Maybe colds were indeed more serious in the past. Maybe aches and pains were so commonplace that few people took notice of a cold that wasn't serious enough to incapacitate them. Brontë, after all, needed Lockwood in a condition where he had to shut up and listen to Nelly Dean's story. A cold that left him on his feet would not have worked as a plot point.

The colds of the past

Like Hercule Poirot or Miss Marple on the trail of a murderer, delving into the history and prehistory of colds involves reconstructing the past from clues available in the present. However, I can't offer as satisfying a denouement as Agatha Christie constructed for her creations. The evidence is simply too thin.

We know that adenoviruses, and probably only adenoviruses, were causing colds among our hunter-gatherer ancestors between 30,000 and 35,000 years ago. We can surmise that most modern cold viruses must have emerged in the last 15,000 years. We can be reasonably confident they were commonplace by the mid-nineteenth century. That leaves an enormous gap between the formation of the first agricultural societies and Emily Brontë sitting down at her writing desk in which we have no real idea how frequent colds were.

We can infer one key point: of the more than 200 cold viruses in circulation today, most are relative newcomers to humanity. A virus with as little time to find a new host as a typical cold virus, meaning any cold virus other than an adenovirus, could only have existed for a little over ten thousand of the hundreds of thousands of years of human history. When our ancestors took to planting crops and feeding animals before slaughtering them, they had no idea that they were building a society in which cold viruses could thrive.

In the millennia since, human society has developed in a way that could have been shaped around the needs of a cold virus. To understand why, we need to delve into the question that lies at the centre of one of the great misunderstandings in medical history: how does a cold virus transmit from one person to another?

Chapter 3

How are colds transmitted?

In 1946, the British Medical Research Council's director, Charles Harington, summoned two scientists. James Lovelock and Owen Lidwell duly travelled to the Medical Research Council's premier institute, the National Institute for Medical Research in the London suburb of Hampstead, and trooped into Harington's office.

Harington asked them, 'Can you go down to Harvard Hospital next week?'[67]

It wasn't really a question. Lovelock and Lidwell knew they were being told to dig out their packing cases and go to Wiltshire.

Neither Lovelock nor Lidwell recorded their reply, but they probably didn't need an explanation. The Medical Research Council, or MRC as the British government's medical research arm is usually known, was a very different beast to the bureaucratic behemoth it has become today. It was small enough that Lovelock and Lidwell had probably heard about the new cold research institute being put together in Wiltshire.

Neither could know they were being set on the path to seminal discoveries in how colds are transmitted from one person to another.

The institute was the brainchild of Christopher Andrewes, who had galvanised cold research a decade earlier when he co-discovered the influenza virus*. He'd since become one of the world's leading infectious disease researchers, giving him enough clout to persuade Harington that the MRC should be coming to grips with the common cold and that the recently abandoned Harvard Hospital was the perfect site to do so.

---

* Andrewes's role in the discovery of the influenza virus is described in Chapter 6.

Harvard Hospital came into being through a 1940 initiative of the American Red Cross and Harvard Medical School, which is how a cluster of prefabricated huts on Harnham Hill, just outside Salisbury and a few miles from Stonehenge, bore the name of a Massachusetts university. Its founders expected the new war to bring a repeat of the devastating 1918 influenza pandemic and they believed the British medical services would need all the help they could get. No pandemic materialised but during the Blitz, British medical services certainly appreciated the help of the American Red Cross who used Harvard Hospital as their base. When the USA joined the war, the US army took over Harvard Hospital as its major blood transfusion centre in Europe. The war's end saw jubilant American soldiers return home by the shipload and Harvard Hospital was one of many abandoned military installations speckling the map of Europe.

Nicely for honeymooners

In February 1946, Andrewes proposed that Harvard Hospital could be turned into a combination of holiday camp and research centre, offering a free holiday in the Wiltshire countryside to anyone who would volunteer for the experiments he and his colleagues would dream up. Harington agreed and commissioned the work needed to turn Harvard Hospital into the Common Cold Unit, usually known as the CCU, although the site retained the name of Harvard Hospital as a nod to the American Red Cross's wartime support.

Harvard Hospital gave the CCU a physical location. Now it needed its human element, which would come in two flavours: a team of researchers and a constant flow of volunteers for them to research. Andrewes wanted a new batch of volunteers every week, enabling a constant source of study subjects for the unit's experiments. His recruitment campaign began with a press conference in which he painted a glowing picture of an all expenses paid fortnight in the English countryside which, thanks to post-war austerity, was not a hard sell despite the strong chance of spending much of that fortnight with a cold.

Andrewes had a puckish sense of humour and, on the spur of the moment, he quipped that the CCU 'would do nicely for honeymooners'[68]. Inevitably, that was the quote that dominated the following day's reporting and it landed the CCU with a reputation it never fully shook off.

That press conference began an ongoing publicity campaign that brought a new batch of volunteers to Harnham Hill every week and made the CCU the butt of many a joke told in music halls, on the radio and in its later years, on television.

On 17 July 1946, a mere five months after Andrewes presented his plans to Harington, the first volunteers were picked up from Salisbury railway station. It's a rate of progress that would make a modern researcher weep. Today, a researcher who submits a proposal in February will still be waiting for committees to meet by July.

Those first volunteers were the first of many. The CCU's publicity campaign delivered an unbroken stream of volunteers from the summer of 1946 until its closure in 1990. Throughout that time, it never shook off its reputation as the place where one could have a free honeymoon if one were willing to risk a cold although, Andrewes later wrote, very few volunteers were newlyweds[69].

Lovelock and Lidwell were among the first researchers who would work there during the CCU's 44 years of operation. Lidwell became interested in the epidemiology of cold transmission through communities which we'll return to in Chapter 5. Lovelock was more interested in the mechanics of how a cold virus found its way from one individual to another.

They couldn't know it, but they were wading into a field that was about to blow up into a decades-long controversy on the other side of the Atlantic. The reverberations of that controversy continue to plague public health policy today.

The dawn of aerosol physics

The question of how a person catches a disease has been posed for as long as it's been known that diseases can be caught, and the answers have sometimes had as much to do with fashion than evidence.

In mid-nineteenth-century Europe, the prevailing view was that diseases were caused by miasma, meaning bad air. By the late 1880s, it was largely accepted that diseases believed to have been caused by miasmas were in fact carried by microbes. Many microbes were associated with nasty smells but the diseases were caused by the microbes, not the smell. Cholera and typhoid, for example, were often transmitted through open sewers which invariably stank. The misconception that it was the smell itself that caused the disease was wrong even though it was logical.

Unfortunately, an important baby was thrown out with the foul-smelling bathwater of miasma theory. As miasma theory went down the drain of medical history along with witchcraft and choleric excess, the concept of airborne disease went with it.

By the early twentieth century, focusing on clean water, clean objects and clean hands dominated discussions of disease control. Cleanliness made intuitive sense to scientists who were used to seeing microbes under microscope slides or on culture plates rather than flying around in the air. Disease-causing microbes can indeed survive in water, on objects and on unwashed hands. The problem was that while those early medical microbiologists were focusing on surfaces and skin, they did not consider the possibility that microbes could be carried through the air as readily as the debunked miasmas.

In the early 1930s, one man wondered whether airborne infection might be compatible with the microbial theory of disease. William Wells was not a medical scientist but a sanitary engineer at Harvard University whose main achievement was an innovation in oyster farming[70]. It wasn't an obvious qualification for tackling the problem of disease transmission, but Wells's background in physics gave him a much deeper understanding of what happens to a sneezed-out droplet than the medical scientists of his day.

If someone could catch an infection by being splattered by droplets from a sneeze, Wells reasoned, then it made sense that one could also catch an infection from a droplet small enough to be inhaled directly – but only if people exhale droplets small enough that they stay in the air instead of falling straight to the ground. He started by working out how small a droplet would have to be to stay in the air for a reasonable length of time.

His calculations revealed that it wouldn't need to be that small. A droplet 100 micrometres* across would take a full six minutes to fall two metres. Wells published his calculations in 1934[71], showing that if someone exhaled a cloud of such droplets into a room, they would float around the room for several minutes. That gave anyone else in the room

---

* A micrometre is a thousandth of a millimetre.

plenty of time to inhale them. In practice, 100 micrometres is a fairly average diameter for a respiratory droplet and we exhale many smaller droplets[72] that float around for much longer than Wells's six minutes.

Wells wasn't finished with his calculations. The six minutes that a 100-micrometre droplet would take to hit the ground depended on it remaining a 100-micrometre droplet. In practice, a lot of water can evaporate off the surface of a droplet in six minutes and as it shrinks, it falls more slowly and becomes more likely to be swept around by air movement.

Wells calculated that a person in a closed room will exhale a cloud of respiratory droplets that floats around a room for an hour. If that person is coughing, sneezing or simply talking, those droplets will leave their nose and mouth at high speed and travel further in their first few seconds than if they were simply exhaled[73]. However, exhalation is all a cold-infected person needs to do to fill a closed room with an invisible cloud of virus-laden droplets.

Wells's calculations were foundational to what would become the discipline of aerosol physics but he didn't feel that he'd answered the question he started with. His training and experience lay in the practicalities of engineering rather than the more theoretical pursuits of physics. He wouldn't trust his conclusions until he'd proved them by experiment.

One of his later collaborators called him an 'eccentric genius'[74], a title that Wells lived up to when he cobbled together a contraption that sprayed fine droplets into the air. He found a way to load those droplets with the recently discovered influenza virus and found that an hour later, the droplet-laden air could infect a ferret[75].

When he published his experiment in 1936, it was by far the most rigorous demonstration of airborne infection to date, but Wells wasn't finished. It was one thing to invent a device for creating disease-carrying droplets in a laboratory. It was quite another to establish that airborne infection happened to human beings in the real world.

In 1937, Wells left Harvard for Philadelphia's University of Pennsylvania, where he worked in the same department as his wife. Mildred Wells was a physician and a medical researcher whose expertise complemented her husband's engineering skills. Between them, they went about proving that infections could be airborne.

The Germantown experiments

In the 1930s, infections were so rife among schoolchildren that schools in well-to-do communities often arranged for daily visits from a doctor. One such school was the Germantown Friends School, which served the mostly Quaker community of Philadelphia's Germantown suburb. The visiting doctor was Theodore Wilder, a colleague of the Wellses at the University of Pennsylvania, who took the opportunity to describe the illnesses plaguing the schoolchildren he tended to.

To a modern reader, Wilder's report reads like a record of unrelenting and overlapping plagues of mumps, measles, chickenpox, whooping cough and scarlet fever, though he didn't neglect to record lesser maladies like upset stomachs, tonsillitis and inevitably colds[76].

To Wilder's readers in the 1930s, it was simply a description of what happened to children in a school.

Wilder noted that nearly all those infections were more common in the winter, which made sense to William Wells. Teachers closed the doors and windows in winter and opened them in the spring which, he calculated, was the difference between completely changing a classroom's air ten times every hour and changing it a hundred times every hour[77]. If the microbes causing those diseases were airborne, they would linger in the classroom for much longer in the winter than in the spring.

The school's poor ventilation offered an opportunity to test Wells's hypothesis that infections were airborne. If they were, then disinfecting the air in some classrooms would lead to a lower rate of infections in those classrooms than in classrooms in which no disinfection was done.

At the time, the only practical way to disinfect air was by using ultraviolet lamps, which can damage the eyes of anyone who looks at them directly[78]. The idea was to interfere with infections, not blind the children, so the Wellses fitted the lamps with baffles that made it impossible for anyone to look at them directly. The baffles also ensured that the ultraviolet could not interfere with microbes on hands or surfaces. If infections were transmitted by children touching each other, touching the same object, or through coughing or sneezing droplets large enough to splatter directly onto each other or objects they might touch, the ultraviolet would do nothing to prevent it.

However, any droplet small enough to linger in the air would be carried into the upper part of the room to be zapped by the ultraviolet. That's because simply by sitting at their desks, children caused the air of their classrooms to circulate between the lower and upper parts of the room. Their 37°C body temperature was warmer than the air in the room, especially in the winter, so a class of children would heat the air close to ground level. The warmed air would rise toward the ceiling where, being further away from the heating effects of small children, it would cool and sink to the lower part of the room. The ongoing vertical circulation wasn't fast enough for anyone to notice it, but it was fast enough to carry any floating droplets up and into the irradiated zone above the baffles.

The Wellses called their approach 'radiant disinfection' although today, such an approach is usually called 'upper room ultraviolet'[79].

In the autumn of 1937, some of the children at the Germantown Friends School started their new year in radiantly disinfected classrooms. It was the beginning of a four-year experiment that would expand to cover several other schools and fortuitously, for the Wellses' results if not the children, those four years encompassed the worst measles outbreak that Philadelphia had ever seen. Having a lot of an easily identifiable disease around made it easy to see if the radiant disinfection made any difference to how many children caught it.

The Germantown experiments yielded a clear result: radiant disinfection roughly halved the number of measles and chickenpox cases[80], which are the most infectious of the infectious diseases[81]. It made little difference to colds or other diseases, but then radiant disinfection of classrooms was never going to prevent all infections. Communal areas in the school, like music rooms and gyms, were not disinfected and many children from the disinfected classes went home with siblings from classrooms that were not disinfected.

The Wellses had started with the question of whether infectious microbes *could* be airborne and they only had to show that some diseases *were* airborne to answer it. They showed that it had been a mistake to dismiss airborne transmission along with miasma theory. They could reasonably expect the debate to move on to the question of *which* diseases were airborne and which were not.

However, not everyone was ready to revise their beliefs about disease transmission.

Sneeze

The five-micrometre doctrine

Sometimes, the lessons of a seminal experiment are immediately embraced into the prevailing wisdom. Sometimes, the prevailing wisdom has too much inertia for one experiment to change it, even if that experiment comprehensively disproves it. The prevailing wisdom that infection could not be airborne was championed by one of the most influential epidemiologists of the mid-twentieth century, giving it what would turn out to be far too much inertia.

Alexander Langmuir played a central role in developing the techniques of epidemiological surveillance, which public health agencies still use to monitor and respond to diseases in human populations. Countless millions of people owe their health to Langmuir. However, he had certain blind spots in his thinking.

Langmuir firmly believed that a disease could only be understood by observing what happened in human populations[82] which, as far as it went, was a reasonable position to take. It was the same view that led William Wells into the schools of Germantown. Langmuir, however, had a much narrower interpretation of what counted as valid observation. He regarded any experimental intervention as suspect even if, like the Wellses' radiant disinfection experiments, it was done in a population that was otherwise behaving as it would in the absence of the intervention. Moreover, Langmuir's disbelief in any such thing as airborne transmission sometimes bordered on zealotry.

Langmuir laid out his views in a 1951 treatise, not on natural infection but on biological warfare[83]. The Cold War was gathering momentum and as director of epidemiology at the National Communicable Disease Center, usually abbreviated to CDC*, part of his job was to worry about America's enemies using diseases as weapons.

The worst-case scenario was an attack with an airborne infectious disease, which could spread faster and would be harder to contain than a disease transmitted any other way. Not to worry, Langmuir argued reassuringly, that could not happen. Granted, there had been reports

---

* The US government's public health agency has had several different names, all of which can be abbreviated to 'CDC'. It is currently the Centers for Disease Control and Prevention.

of airborne diseases but those had all been in laboratory experiments. Scientists could make a lot of things happen in laboratories that didn't happen anywhere else. Mechanically generating a cloud of fine droplets, as scientists like Wells had done, did not prove that people went around breathing microbes into the air[84].

Langmuir must have known about the Germantown experiments because a few years earlier, he and William Wells had both sat on a committee on airborne infection control. As the committee's secretary, it had fallen to Langmuir to draft the committee's report that stated that the evidence for airborne infection was inconclusive[85]. He later acknowledged that the committee had 'struggled for a long time'[86] over that conclusion, although he gave no specifics. There's no way to know how much of the struggling involved arguments between Langmuir and Wells, whose views of the committee's 'consensus' are not on record.

The Wellses continued with their own writings that treated airborne infection as established fact. Langmuir never mentioned the Germantown experiments in anything he wrote while he worked for the CDC.

Buried in the many details of Langmuir's 1951 report on biological warfare was one critical detail that slipped into the prevailing wisdom without anyone noticing. It stated that, 'Particles larger than 5 microns* in diameter are almost completely removed in the nose and upper respiratory passages'[87], which he presented as definitive evidence that infections couldn't be airborne.

It was a sentence that would cause decades of misconception.

The sentence was factually correct. The various respiratory passages between the nose and the lungs do indeed filter inhaled air so effectively that nothing larger than five micrometres across can get down to the bronchioles and the lungs[88]. That's good news as far as it goes; infection in the bronchioles and lungs is a lot more serious than infection of the upper respiratory tract. Microbes like the cold viruses do not need to get through the respiratory passages to infect someone. They can find plenty of cells to infect above those filtering mechanisms, in the nose and the throat. A cold virus doesn't need to be in one of those tiny five-micrometre droplets to reach the upper respiratory tract and, as Wells

---

* 'Micron' is an abbreviation for 'micrometre', although the more commonly used abbreviation today is 'μm'.

had calculated back in 1934, droplets much larger than five micrometres can stay airborne long enough for someone to breathe them in.

Such was Langmuir's influence in the field of public health that his views shaped doctrine. Airborne transmission was negligible, Langmuir's doctrine stated, because it depended on those very rare droplets of five micrometres or less. It's a measure of Langmuir's influence that long after Wells showed both that five-micrometre droplets are not particularly rare and that much larger droplets can stay airborne, nobody seems to have challenged Langmuir's premise that airborne infection depended on droplets getting down the windpipe and into the lungs.

Under better circumstances, the Wellses might have mounted such a challenge, but Mildred died soon after the report was published. William became the sole carer for their intellectually disabled son[89], limiting his time for research and advocacy. Things got even worse when Wells collapsed in his home. Unable to reach the phone or to explain to his son how to use it, he lay helpless for two days before someone found him and got him to a hospital. That intervention saved his life but he was left paralysed from the waist down by a cancer that had metastasised into his spinal column.

The effects of the cancer were not purely physical. The man whose meticulous calculations were foundational to modern aerosol science began to lose his grip on reality. He was then involved in a study on the bacterial disease tuberculosis and a colleague later described how study meetings would lurch between Wells's characteristic insights and 'frankly psychotic'[90] outbursts about some hidden enemy torturing him with electric shocks to his legs. Nevertheless, he remained a leading scientist on the study that conclusively demonstrated airborne transmission of tuberculosis which was published in 1962[91], a year before Wells's death.

Langmuir remained unpersuaded.

Airborne, contact or both?

When James Lovelock arrived at the CCU in 1946, he must have been aware of the Wellses' experiments, but Langmuir's vocal opposition still lay in the future. The idea that some infectious diseases were airborne probably didn't seem as contentious as it would be a decade later although nobody had looked into whether colds might be among them.

Lovelock probably had mixed feelings about his new posting. During the war years, he'd been exempt from military service as a registered conscientious objector and had worked on whatever projects the MRC had needed him to work on.

He first got involved in cold research during the Luftwaffe's night-bombing Blitz, when crowded and poorly ventilated air raid shelters raised the same fears of pandemic influenza that lay behind the founding of Harvard Hospital. The MRC convened an informal Air Hygiene Unit which included Lovelock and Lidwell. No pandemic materialised but crowded, poorly ventilated air raid shelters, where the air quality was often so poor that it was impossible to strike a match, were hubs for airborne infections like colds.

Air hygiene was one of many projects that came Lovelock's way. Another such project was to use shaved rabbits to assess the effect of a near miss from a flamethrower. Lovelock couldn't stomach the idea of scorching a rabbit so he allowed his own skin to be seared instead[92].

Another project stemmed from a request for help to contain colds on a US Army Air Force bomber base. The USAAF didn't think colds were a trivial problem. Anyone who has taken a commercial flight while suffering from a cold will know how uncomfortable it makes a cabin pressurised to no more than 8,000 feet above sea level[93]. In an unpressurised aircraft flying at 30,000 feet, a cold is agonising[94].

The end of the war brought an end to both the military projects and to the fear of a pandemic. Lovelock and Lidwell were transferred from the National Institute of Medical Research to the London School of Hygiene and Tropical Medicine without a clear role. They probably felt like the MRC's middle children until Harington sent them to the embryonic CCU.

Lovelock didn't record his first impressions of the CCU but they probably weren't favourable. Harvard Hospital's prefabricated huts weren't designed to make a good first impression and having been given a week to uproot his young family from London, Lovelock probably wasn't in the most favourable frame of mind. One Dutch researcher later likened the rows of drab huts to a Nazi concentration camp in which he'd been imprisoned[95] although the CCU's founder and director, Christopher Andrewes, offered a cheerier description, believing the twin water towers looked like the funnels of the luxury liner *Mauretania*[96].

On the upside, Lovelock had been struggling to make ends meet in London and his new post came with accommodation in one of Harvard Hospital's huts. It relieved his financial pressures and in their first winter, the Lovelock family very much appreciated that the hospital's American designers had fitted their hut with an efficient heating system[97], which was something of a novelty in 1940s Britain.

Lovelock's research picked up where the Wellses' had left off, albeit with a much greater focus on colds. However, Lovelock approached the question from the opposite direction. Where Wells had started by asking whether viruses could plausibly be carried by airborne droplets, Lovelock started by asking whether colds could be transmitted by direct contact. It was assumed that a virus could be carried out of one person's mouth onto some object and then that someone else's hand would carry it off that object and into their own mouth. It was an assumption that involved multiple untested steps which, to Lovelock, made it a fragile enough assumption to be worth examining in detail.

Lovelock started by cobbling together a device that dripped fluorescent dye through a tube running next to someone's nose. Then he inveigled some of his colleagues to wear his device while they played a card game. At the end of the game, a sweep of ultraviolet light revealed the dye all over the cards, the table and everyone's fingers[98].

He'd proved that fluid dripping out of a cold-infested nose could get as far as someone else's fingers but that left the question of whether those fingers were likely to put it up their own noses. Lovelock answered that question by travelling on the London Underground and surreptitiously recording how often his fellow passengers touched their faces. The answer, it turned out was very often[99].

The two experiments showed that cold viruses *could* conceivably travel out of one person's nose and into another's via objects and hands but that was not the same as proving that cold viruses *did* transmit by that route. Nor had Lovelock ruled out airborne transmission, which had been central to the concerns about influenza in air raid shelters that had got him started on the question of cold transmission in the first place.

Like Wells, Lovelock sought to answer those questions with a practical experiment. He invited people who had colds, mostly children from Salisbury, to the CCU and experimented with different ways they might give their colds to uninfected adult volunteers. First, the cold-carriers were ushered into a hut in which they were separated from uninfected volunteers by hanging blankets from a rope. For two hours, the uninfected volunteers read books or sewed while the cold-carriers were encouraged to shout and sing and, at one point, were given sneezing powder. The experiment ensured that the cold-carriers were spraying cold viruses all over their side of the room but the only way those viruses could get to the uninfected volunteers was by floating over the blankets.

The cold-carriers were then taken to lunch and a card game with a second group of uninfected adult volunteers, exposing those volunteers to both airborne and contact infection.

The room was aired to remove any airborne infection and a third group of uninfected adult volunteers was brought in to spend a couple of hours playing cards with the same cards over the same table, exposing them to contact infection only.

In the event, very few volunteers in any of the three groups caught colds. Only two of the 25 who had been exposed to airborne transmission caught colds, three among the 32 who had shared lunch and a card game with the carriers, and another two among the 25 who had shared lunch and a card game with the carriers[100].

Good news for most of the volunteers whose time at the CCU was unblighted by a cold. Bad news for Lovelock as he tried to interpret his results.

His main conclusion was that it's not easy to give an adult volunteer a cold which, given what we know now, is not particularly surprising. Adults don't get colds as often as children[101], mostly because the older someone gets, the more cold viruses they are exposed to and the more to which they develop at least some immunity. In 1952, Lovelock and his colleagues knew that children catch more colds than adults, but they were only just beginning to come to terms with the enormous diversity of the cold viruses. He might have infected more volunteers if he'd stuck to adult cold-carriers because even if he didn't know which virus he was dealing with, he'd have known it could infect at least some adults. In the event, there simply wasn't enough known about cold viruses for Lovelock to consider such an approach.

Nevertheless, Lovelock was able to draw one clear conclusion: that even a few volunteers from all three groups were infected showed that both airborne and contact transmission were at least possible.

Lovelock's card games and blanket draping were foundational in our understanding of how colds are transmitted and they supported the Wellses' earlier work in building a counterpoint to Langmuir's anti-airborne dogmatism.

Lovelock left the CCU soon after his transmission experiments, which became little more than a prologue to his more celebrated later career. They got him thinking about how he might sample virus-laden droplets directly from the air, which would be a much more conclusive demonstration of airborne transmission than seeing whether viruses could get over a draped blanket. He didn't succeed, and even today, waiting for an experimental animal or a human volunteer to be infected remains the only way to test for airborne virus.

However, it got Lovelock thinking more broadly about what might be floating around and led him to invent a device that can detect trace quantities of gaseous chemicals in the air. He went on to use his electron capture detector, as he called it, to show that the atmosphere is suffused with trace quantities of chlorofluorocarbons, or CFCs. His work underpinned the international agreements to stop using CFCs before they destroy the atmospheric ozone layer that protects the earth's surface from ultraviolet light[102]. That in turn got him thinking about the interactions between the physical environment of ocean and atmosphere with the biological world of the organisms that change their composition by living in them. His train of thought led him to formulate the Gaia hypothesis, making him one of the most famous scientists of the late twentieth century and sparking a much higher-profile controversy than that of whether disease can be airborne or not.

Meanwhile, his experiment remained the last word on whether colds can be airborne for more than a decade.

Card games

In 1967, the question of cold transmission was picked up by Jack Gwaltney at the University of Virginia, who would later carry out the seminal research on how a misdirected immune response turns a virus infection into a cold*. Virology had come a long way in the 15 years

---

* As described in Chapter 1.

since Lovelock was recruiting sniffling children in Salisbury. Instead of simply exposing volunteers without colds to sniffling volunteers, Gwaltney and his colleague, Owen Hendley, were able to base their experiments around one selected virus and to assess their volunteers' antibodies to ensure they were not immune to it.

They chose one of the rhinoviruses, a group of more than 100 different viruses that cause between a third and a half of all colds[103]. Gwaltney and Hendley started by dripping fluid containing rhinoviruses onto surfaces, then getting volunteers to touch the surfaces and confirming that they had a live virus on their fingers[104]. Then they got twelve volunteers to contaminate their hands with rhinovirus and another twelve to hold their hands and then pick their noses. Eleven of those twelve developed colds[105].

They had proved that contact transmission was a viable route up someone's nose, but they hadn't ruled out airborne transmission. They went on to infect volunteers with their rhinovirus and, if those volunteers went on to develop a cold, placed them at one end of a room while introducing a second group of volunteers to the other side. The two groups of volunteers were separated by a mesh that ensured the only way the rhinovirus could travel from the cold-carriers to the uninfected volunteers was by floating from one end of the room to the other. Not one of the healthy volunteers was infected.

Eleven out of twelve for contact transmission and zero out of ten for airborne looked like a firm win for contact, but that wasn't the end of the story. A few years later, Elliot Dick at the University of Wisconsin ran another set of experiments with a rhinovirus and got very different results.

Dick's experiment started by deliberately infecting men with a rhinovirus and when they were coughing and sneezing like champions, he sat them down with another dozen men for a twelve-hour poker game.

Like Lovelock, Hendley and Gwaltney, Dick designed his experiment to expose some people to airborne transmission, some to contact transmission and some to both. Like Lovelock, he based his experiment around a card game because every player would touch the cards and, presumably, because it was the only way of persuading people to devote hours at a time to passing around articles that might carry rhinovirus.

It's hard to imagine anyone volunteering to play pass the parcel for 12 hours.

The poker game exposed volunteers to both airborne and contact transmission. If a cold-carrier had any virus on their hands, they would transfer it to everyone else's hands via the cards. As Lovelock established on the London Underground, everybody touches their face sooner or later – sooner, in most cases – giving a rhinovirus a clear route out of a carrier's nose and up an uninfected volunteer's. At the same time, the cold-carriers were exhaling rhinovirus into the room and presumably firing it across the table at each other by coughing and sneezing.

To separate the contact and airborne routes, Dick fitted some of the volunteers with contrivances that prevented them from touching their faces. Some wore wide plastic collars while others wore rigid arm braces. Dick did not record whether the indignity of being collared like a dog recovering from an operation affected anybody's poker game, but the restraints did their job: if a player couldn't touch their face and washed their hands thoroughly before meal breaks, contact transmission was blocked and the restrained volunteers could only be infected if the rhinovirus was airborne.

Dick's poker marathons started at eight o'clock in the morning and, allowing breaks for meals and sampling, went on until eleven at night. By then, the volunteers must have been exhausted but the experiments weren't finished.

The chips, pencils, furniture and the cards, which were by now 'soggy'[106], were taken into another room where another dozen uninfected volunteers sat down for another twelve-hour poker game. That second group handled everything the cold-carriers had touched but they breathed none of the air the cold-carriers had breathed.

Where Gwaltney's and Hendley's results had supported contact transmission, Dick's came down firmly on the side of airborne transmission: among the volunteers who played cards with the cold-carriers, ten of the eighteen who had worn restraints caught the rhinovirus and twelve of the eighteen who were unrestrained. However, the soggy cards infected none of the 12 players in the second card game[107].

The contradiction between Dick's results in Wisconsin and Gwaltney's and Hendley's in Virginia presents something of a conundrum. It's not likely that rhinoviruses' preference for contact or airborne transmission depends on which side of the Mason-Dixon Line they're transmitting on. The different conclusions may be down to differences in the experimental setups or they may be because the two experiments used different rhinoviruses. It's possible that some rhinoviruses transmit mostly by the contact route while others are mostly airborne. We don't know because, since Dick published his results in 1987, there has been very little experimental work on the airborne transmission of rhinoviruses or, with the recent exception of SARS-CoV-2, of cold viruses in general.

However, there is one conclusion that we can draw by combining Lovelock's early work at the CCU with Gwaltney's and Hendley's experiments and finally with Dick's: to ask whether colds are transmitted by the contact or the airborne route is to present a false dichotomy. Certain viruses or certain situations may favour one route over the other but there is ample evidence that cold viruses can transmit by both routes.

Part of the reason for the lack of subsequent experiments is that by the 1980s, airborne transmission was no longer as controversial as it had been when Dick orchestrated those card games two decades earlier. The tuberculosis study that had been William Wells's swansong was one of several showing airborne transmission in several infectious diseases. Even Alexander Langmuir relented, albeit long after he retired from the CDC in 1970. In 1980, he finally acknowledged that some diseases spread at a rate too fast to be explained by anything but airborne transmission. He still gave no credit to the Wellses' direct demonstration of airborne transmission.

Langmuir withdrew from the debate with an admonition that 'the epidemiologists of the future should clearly practice their profession with a greater humility and skepticism of their past teachings than some, at least one, epidemiologist of my close personal acquaintance'[108].

It was as close as he ever came to admitting he'd been on the wrong side of the debate for three decades.

The Langmuir legacy

By the end of the twentieth century, epidemiologists and virologists working on colds accepted that they could be transmitted by both the airborne and contact routes. The question was no longer whether either route did or did not happen, but which route was more important for a given cold virus under a given set of circumstances.

That question came to the fore in 2002, driven by an epidemic of a disease called severe acute respiratory syndrome, better known as SARS, a disease caused by a virus very closely related to some of the cold viruses*. The epidemic led a cross-disciplinary team of Hong Kong University scientists to revisit the calculations William Wells published back in 1934. With the benefit of 70 years of advances in physics and aerosol sciences, they confirmed that Wells had got it right[109].

Their result that would have given Wells something to smile about but by then, it wasn't challenging any paradigms – or so most medical researchers who were interested in cold viruses and their relatives thought.

They had a rude shock on the last Saturday of March 2020 when the WHO took to Twitter† to announce 'FACT: #COVID19 is NOT airborne'[110].

The tweet went on to explain that if we all kept a metre between us, we couldn't splatter each other with any virus-laden droplets we might cough and sneeze in each other's direction. It also recommended disinfecting surfaces and regular handwashing to break the contact transmission route which, the WHO appeared to believe, was all that was needed to protect against a 'NOT airborne' virus.

The tweet did not age well. It was soon painfully obvious that COVID-19 was indeed airborne, although it took some time for the WHO to change its position.

A week after that tweet, a group of 36 experts in the movements of airborne droplets Zoomed with the panel of experts hastily convened to advise the WHO on COVID-19[111]. In the course of the conversation, some of those aerosol scientists realised that they were at cross purposes with the WHO's advisors over a critical point: the WHO advisors believed that the only way a virus could be airborne, in any meaningful sense of the word, was if it was contained in a droplet with a diameter of five micrometres or less. The WHO advisors believed that any particle larger than five micrometres would simply fall to the ground.

---

* Chapter 9 describes the SARS epidemic and the virus that causes it in more detail, including that virus's close relationship with SARS-CoV-2 and the endemic coronaviruses.

† Since renamed X, a name it retains at the time of writing.

It was a view that perplexed the aerosol scientists. Airborne pollen, which can drift for miles to fertilise another plant or block a nose with hay fever, is usually between 20 and 50 micrometres in diameter[112]. When I asked one of the aerosol scientists, Jose-Luis Jimenez of the University of Colorado Boulder, about it, he described the WHO advisors' view as, 'just mind-boggling because I knew that particles much higher than five microns cross the Atlantic'[113].

Lidia Morawska, perhaps the world's foremost expert on the physics of airborne droplets, tried to explain the long-established physics that proved them wrong, but she didn't get very far.

Jimenez recalled how the chair 'started yelling at her', shouting her down to tell her she was wrong 'with a level of anger that, I mean, left us shocked.'[114]

Langmuir's doctrine, already disproven when he wrote it in 1951, was alive and unhelpfully well in 2020.

The aerosol scientists left the meeting feeling dismayed and frustrated. Many of their epidemiologist colleagues were equally mystified. It was already known that SARS-CoV-2 was at least as transmissible as a pandemic influenza virus[115] which would have persuaded even Langmuir that it was airborne.

How, the aerosol scientists wondered, was a panel of WHO experts still insisting there was no such thing as airborne transmission 40 years after the concept's greatest opponent acknowledged that there was?

Katherine Randall investigates

That question came up in many a bemused conversation after that Zoom call, until it landed on the desk of Katherine Randall, a postgraduate student at Virginia Polytechnic Institute and State University*, whose doctoral research had been brought to a shuddering halt by the COVID-19 lockdowns.

Randall had never planned to investigate how different disciplines had ended up with contradictory views of whether infections can be airborne but in 2020, it was a research question with one cardinal virtue: the answer lay in documents that had all been digitised and

---

* Usually known as Virginia Tech.

made available online. With libraries closed and in-person interviews now unacceptably risky, Randall needed something she could take on with only a laptop and an internet connection. She embarked on what one journalist later called 'scholastic detective work'[116], likening it to the forensic analysis so beloved of TV murder mysteries. Randall later said that the 'comparison to blood sprays is … the coolest way anyone has ever described scrolling through the CDC archives at one a.m. on a random Wednesday in June'[117].

Randall uncovered the five-micrometre misunderstanding at the heart of the objection to airborne transmission[118] that continued to influence infection control long after epidemiologists and aerosol scientists had accepted that some infections were airborne. Langmuir's doctrine had gained such inertia that it transcended Langmuir's own refutation of it. By then, many of the people who regarded it as a truism had trained after Langmuir's retirement and would not have recognised his name. Somehow, nobody had noticed that related disciplines were following contradictory doctrines before that frustrating Zoom call[119].

Not every country followed the WHO's lead in adhering to the five-micrometre doctrine. Some demurred and designed their responses around preventing airborne transmission.

Among the demurrers was Hong Kong, where public health policymakers drew on their experience with the SARS epidemic. As early as January 2020, well before the danger posed by the virus that was still several weeks from being named SARS-CoV-2 was understood, Hong Kong care home staff were required to wear surgical facemasks to prevent airborne transmission*.

Meanwhile in Britain, policymakers followed the WHO's position, emphasising physical distance and handwashing to prevent contact transmission. Care homes were neither advised to issue facemasks nor to do anything else to prevent airborne transmission. The consequences amounted to a stark lesson for Langmuir's remaining adherents: between March and June 2020, COVID-19 killed more than 10,000 care home residents in Britain but hardly any in Hong Kong[120].

---

* Chapter 13 takes a deeper dive into the mechanisms and effectiveness of different facemasks.

By the second half of 2020, the WHO acknowledged that airborne transmission was at least a possibility. The WHO's chief scientist later stated that the organisational failure to recognise airborne transmission was the greatest regret of her five-year tenure. Even after ventilation and facemasks were included in the WHO's advice for mitigation, she acknowledged, 'we were not forcefully saying: "this is an airborne virus". I regret that we didn't do this much, much earlier'[121].

Decades after there had been ample evidence that infections can be airborne and that fast-spreading infections probably *are* airborne, that evidence finally appears to have found traction among public health policymakers. There is no way to know how many lives could have been saved if it had found that acknowledgement at the beginning of the COVID-19 pandemic.

As I write this, the British press's health sections are abuzz with a worrying rise in measles that has already killed one child, one of the infections that was central to the Wellses' proof of the concept of airborne infection. Not to worry, says a spokesperson for the Early Years Alliance which represents nursery providers; their members are 'using all those good hygiene practices, staff have got [personal protective equipment], they've got aprons, gloves, some settings still keep masks'[122].

It's enough to make an epidemiologist look for a solid object to bang their head against. No amount of aprons, gloves or masked-up staff members 'in some settings', whatever that means, will prevent the airborne transmission of measles between the children. Some 45 years after Langmuir's recantation, his doctrine continues to guide ideas on infection control.

Chapter 4

What makes someone more susceptible to a cold?

A woman crouches on a roof outside a window, enduring the driving rain soaking her hooded cloak and flannelled petticoats to eavesdrop on the hatching of a murderous plot. She's listening to the dastardly Sir Percival Glyde and his friend, Count Fosco – whose foreign accent marks him as even more dastardly – planning to bump off Sir Percival's wife for her money. The details are worth the drenching because the woman being plotted against is her dear half-sister.

Sir Percival and Count Fosco conclude their conversation with appropriately villainous laughter, leaving the drenched woman to retire to her bedroom and record their plans in her diary. There's a price to pay for her sleuthing: the rain has chilled her to the bone. She is writing through shivers and then through the throbbing heat of fever. She weakens until she fears that 'if I lie down now, how do I know that I may have the sense and strength to rise again?'[123] Her writing trails into illegibility.

Marian Halcombe has caught a cold.

It gets worse. Her condition deteriorates into full-blown typhus, rendering her helpless beneath the scheming Sir Percival's roof. What will become of Marian Halcombe?

If Marian's exploits read more like the stuff of melodrama than real life, it's because the action took place in the pages of Wilkie Collins's 1859 novel, *The Woman in White*. It's now considered one of the first modern mystery novels and, more pertinent to an exploration of the common cold, Marian's illness gives us an idea of how Victorians understood colds a couple of decades before anyone realised microbes had anything to do with it.

Chilled and soaked

Marian was rendered quiescent because she got cold and wet, which became the gateway to the potentially lethal – and, we now know, distinctly unlikely – complication of typhus. When I was a child, my mother made that same association between getting chilled and catching a cold when she admonished me, 'Don't go out without your coat, if it rains you'll catch your death of cold' or 'wrap up warm or you'll get pneumonia'.

My mother's views in the twentieth century, and presumably Collins's views in the nineteenth, were reflected in an ethnography of parts of the USA and Mexico published in 2008. A century and a half and an ocean away from the fictional Marian Halcombe, colds were still being blamed on getting chilled and soaked[124].

That ethnography is one of the very few systematic studies on what people believe causes colds. It's a remarkable gap in the literature given that whenever I mention that I'm writing about colds, the conversation turns to what everybody's mother told them they must never do or they would catch their death of cold, flu and, if they were feeling hyperbolic, pneumonia, which is at least a little more likely than typhus.

Given the dearth of published literature, I resorted to a very unscientific straw poll of people I happen to know from around the world. I was expecting to hear a lot of different ideas but in the event, nearly everyone I asked had been told that they would catch a cold by getting chilled and soaked. Even worse than a chilled body was a chilled head; I was told that wet hair is blamed for colds in Russia, Greece, Spain, Turkey, the Philippines and Iran.

There were some variations on that theme: a Chinese correspondent's mother told her that if she got wet hair by being caught in the rain, she could fend it off by immediately washing her hair, while an Indian correspondent had been told that going to bed immediately after oiling her hair was every bit as likely to cause a cold as getting it rained on.

The exceptions were the correspondents from Singapore and Malawi, both of whom told me that obviously, you catch a cold from someone who has a cold. Both seemed bemused to hear that anyone thought getting chilled and soaked had anything to do with it. At first glance, the Malawian and Singaporean views appear to be the most closely aligned with the scientific evidence. Back in the 1920s, Alphonse Dochez showed that colds are caused by cold viruses* and cold viruses are not conjured into existence by wet hair.

My British mother was fully on board with the global belief that getting chilled and soaked would inevitably lead to a cold, but she added something that I didn't hear from any of my international correspondents: she believed that wet feet induced colds as inevitably as wet hair. Her view appears to have been widely shared in Britain, if not beyond, as observed by a North London general practitioner and anthropologist called Cecil Helman. Practising in the 1970s, he found that many of his patients blamed wet feet for colds and firmly believed that getting soggy socks from stepping in a puddle was bound to lead to a headful of catarrh[125].

Helman found that most of his patients thought getting chilled and soaked mattered because they believed that colds were caught through the skin of their hands and faces. Given the century of research showing that colds are caused by viruses getting into the mouth and nose, which are rather important parts of the face, and that they can be inadvertently carried there by the hands, it would be easy to attribute their views to outdated superstition – but perhaps we shouldn't be quite so hasty.

As we saw in Chapter 1, the illness of a cold is caused by the immune system's response to the virus infection rather than the infection itself. Many cold virus infections cause no symptoms which means, to use the parlance of epidemiology, that a virus is a *necessary* element of a cold but is not *sufficient* to cause a cold. If we pretend for a moment that Marian Halcombe's cold-inducing exploits were factual rather than fictional, we can infer that there was a cold virus replicating up her nose but that virus would have been only the first step toward her inconveniently timed cold. If she had been infected with that particular virus before, her lymphocytes and antibodies might have prevented it from infecting her again. If she were susceptible, she would still need to have enough of the common cold constitution for it to make her ill.

The scientists who first described the common cold constitution described it as if it were a fixed characteristic in every person[126], with each of us fated with a constant level of susceptibility to colds. It's an

---

* Chapter 1 describes how Dochez did it.

assumption that begs a question: is it possible that the way someone reacts to a cold virus is not in fact fixed but may be influenced by external factors like, for example, getting a good soaking just at the moment the virus is starting to replicate?

Some medical scientists have taken the view that if there's a plausible mechanism for such a widely believed association between chilling and catching a cold, it would be a mistake to dismiss that belief out of hand.

Among them was, perhaps unsurprisingly, Christopher Andrewes.

The soggy socks experiment

In the late 1940s, the virology of the common cold had not advanced very far since Dochez's experiments two decades earlier. There had been significant progress in the study of influenza, but that progress had shown that the influenza viruses only caused a small minority of colds. The majority were attributed to some presumably virological entity or entities that remained mysterious.

The dearth of information on non-influenzal colds, as the unexplained colds were called, divided most medical researchers into two schools of thought. Some started to wonder if non-influenzal cold viruses existed at all. Perhaps, they came to believe, Alphonse Dochez had been wrong and the widespread belief that colds were caused by getting cold and wet had been right all along.

Others believed that Dochez had been completely right: one caught a cold when one was infected with a cold virus and no further explanation was needed.

The concept of the common cold constitution was introduced in 1947[127], which may have influenced Andrewes when he started wondering if the division between those schools of thought was premised on a false dichotomy[128]. If chilling caused a cold, he pointed out, then Arctic explorers would be constantly afflicted with colds but in fact, they very rarely complained of them. On the other hand, he and his colleagues had been stuffing cold viruses up the noses of CCU volunteers for years and only a minority developed colds, which strongly indicated that there was something more to getting a cold than being infected with a cold virus. He wondered if part of that something might be getting chilled and soaked.

Colds are, after all, more frequent in the winter – or at least, they are in temperate climates like those of Europe and North America which, at that time, were the only places where colds had been studied in detail. The association between cold temperatures and catching colds was so ingrained in the mostly British CCU researchers that they assumed that colds simply did not happen in the tropics. If you're reading this anywhere in the tropics, that view will bemuse you because you'll know that colds sweep through tropical countries as frequently and as annoyingly as they sweep through temperate countries. It would be a couple of decades before the CCU researchers realised that[129].

Andrewes wasn't the first to connect the two meanings of cold. In 1919, Stuart Mudd and Samuel Grant at Washington University in St Louis had persuaded volunteers to sit in a cold room with a cork between their teeth while they stuck a thermometer of their own design into their throats. Mudd and Grant first tried the experiment on dogs, but they soon found that dogs wouldn't sit still with a cork in their teeth without being anaesthetised which, they found, interfered with the reflexes they were trying to measure. They didn't record where they found human volunteers who were more cooperative than dogs, but they did find that while their volunteers maintained a constant core temperature, breathing cold air dropped the temperature in the tissue lining their throats[130].

Those uncomfortable volunteers responded to cold air by closing off the blood vessels in their throats which, Mudd and Grant speculated, might make someone more prone to a microbial infection. More recently, the same thing has been shown to happen in the sinuses, with the blood vessels closing off while mucus production is often increased. That's why so many of us get runny noses when we go out in the cold[131].

For Andrewes, a full understanding of what would later be dubbed 'skier's nose' lay in the future, but Mudd's and Grant's speculation that the body's response to an infection might be affected by its response to chilling seemed at least plausible. Moreover, this was exactly the sort of question he'd conceived the CCU to answer.

Andrewes devised an experiment in which volunteers were divided into three groups of six. The first group was given a warm bath and then left to dry in a cold corridor for as long as they could stand it, and asked to wear wet socks for several hours after that. The British view that cold feet and soggy socks causes colds was alive and well three decades before Helman wrote it down.

A second group was inoculated by dripping a suspension of cold virus up their noses. A third group was given both the cold treatment and the virus inoculation.

Not surprisingly, the soggy socks group 'felt rather chilly and unhappy for a time' but none developed colds. Two of the six inoculated up the nose developed colds while four of the group given both the soggy socks treatment and the virus inoculations developed colds 'of which 2 were rather good ones'[132].

It looked like getting cold did indeed make people more likely to develop a cold once they were exposed to the virus but, as Andrewes put it, 'we were foolish enough to repeat this experiment, with a contrary result'[133]. Soggy socks alone still caused no colds but this time, there were twice as many colds among people who received the cold up the nose without the soggy socks than among those who had both the soggy socks and the cold up the nose.

Despite his characteristic humour, Andrewes must have known that repeating the experiment was the opposite of foolish. Consistently deriving the same results from the same experiment is a cornerstone of the scientific method. Had Andrewes consulted a statistician, he would have been told that his mistake was that he hadn't repeated it enough. Given how few volunteers developed colds under any condition, Andrewes would have needed to test each condition on a lot more than 12 volunteers to see a consistent difference between them.

Reading reports of the CCU's early experiments, it is evident that the scientific team that Andrewes recruited did not include a statistician. The soggy socks experiment was one of many that wasted a good setup by not doing it on enough people to get a result that could be interpreted.

The Chicago chilling experiment

In 1949, Andrewes regaled an audience at Harvard Medical School with the story of the CCU's antics, including the wet socks experiment. A young microbiologist called George Gee Jackson was given the job of driving Andrewes to his various appointments and lectures around Boston. It was, Jackson wrote 40 years later, such a remarkable privilege for a 'scientific nobody'[134] to be alone with a man of Andrewes's standing that it pointed Jackson himself toward common cold research.

Jackson picked up an interest in colds from Andrewes, but he didn't share Andrewes's indifference to statistics. A decade later, when Jackson was becoming a scientific somebody at Chicago's University of Illinois, he recruited 428 volunteers to investigate whether chilling makes someone infected with a cold virus more likely to develop a cold. That was a large enough experiment to deliver a reliable result.

Jackson did not involve soggy socks in his experiment, but instead he placed volunteers in cold rooms to chill them from head to toe. Some were stripped to their underwear and left in a 16°C (61°F) room for four hours. Others were allowed to dress for cold weather and placed in a room at -17°C (1.4°F) for two hours, which is roughly the temperature of a household freezer or a cold winter's day in Chicago. The more fortunate volunteers were spared the shivering, but then half the group had a cold virus dripped up their nose.

Jackson found that around a third of the volunteers he inoculated developed a cold, irrespective of how cold he'd made them[135]. All that shivering did not affect how likely someone was to develop a cold.

When Jackson published his experiment in 1958, it was the most robust test of the chilling hypothesis. It persuaded cold researchers that chilling had nothing to do with catching colds although, as my mother and my correspondents demonstrated, it was not a view that prevailed beyond medical scientists.

There was, however, a result that received no more than a throwaway sentence in Jackson's report: among the people deliberately infected with a cold virus, more women than men went on to develop a cold. They weren't the only ones to notice that difference; the CCU scientists, who inoculated volunteers week-in and week-out for years, found that their inoculations induced colds in slightly more than half of their women volunteers but slightly less than half of the men[136].

If a bunch of people are inoculated with a cold virus and one group is more likely to get a cold than another – whether the difference is based on gender or any other characteristic – then there must be some factor interceding between inoculation and illness that is more active in one group than the other. Neither Andrewes nor Jackson found any evidence that a cold person is more likely to develop a cold, but the difference between men and women showed that there was at least one systematic difference in who was likely to have a common cold constitution and if there was one, there were likely to be others.

Freshers' flu

Jackson's conclusion that chilling was not involved in colds put the issue to bed for more than 40 years, until it was dragged back out by Ronald Eccles at Cardiff University's Common Cold Centre.

Eccles set up the Common Cold Centre in 1989[137], funded by pharmaceutical companies to test cold remedies for pharmacies to sell over the counter. The Centre's clinical trial work gave Eccles a platform to explore the science of colds more broadly[138], and when the CCU closed in 1990, Cardiff's Common Cold Centre became Britain's leading cold research establishment.

As the new millennium dawned, Eccles began to wonder whether Jackson really had proved that chilling does not cause colds. His view was that Jackson's experiment was fundamentally flawed because his technique for artificially inoculating his volunteers was fundamentally different to a natural cold infection.

The most obvious difference was that Jackson's inoculations contained far more cold virus than someone would be likely to breathe in from a cloud of droplets. An infection that started with an unnaturally huge amount of virus might have been such a strong inflammatory stimulus that it overrode any other factor that might be involved in developing a naturally acquired cold.

On the other hand, a persistent frustration among the CCU researchers was that, as one CCU virologist complained to a journalist that, 'believe it or not, one of our biggest problems is giving people colds'[139]. Despite the high concentration of virus in their inoculations, it was only between a third and a fifth of inoculated volunteers who developed a cold.

Eccles's view was that to understand naturally caught colds, he needed to avoid the vagaries of artificial inoculations and study naturally caught colds[140]. That presented a practical problem. Andrewes and Jackson had resorted to artificial inoculation for a reason: cold symptoms don't appear until between one and four days after the initial infection. A virus might be replicating away in someone's throat today that will give them a raging sore throat tomorrow but until then, they look, sound and smell exactly like someone who does not have a cold virus lurking in a single one of their cells.

Artificial inoculation, for all its shortcomings, had enabled Andrewes and Jackson to assess the effect of chilling at precisely the moment in which they were infected. If Eccles needed volunteers in the early stage of a naturally acquired infection, he needed a different approach.

His solution was to recruit a lot of volunteers from a population in which there were a lot of colds and, being based at Cardiff University, Eccles had such a population right on his doorstep. Professors don't need to go around infecting people to turn universities into hotbeds of cold viruses. Students do that without any help. One student breathing cold viruses into a poorly ventilated lecture theatre spreads the virus like wildfire.

When I was an undergraduate, we used to talk ruefully about the triannual 'freshers' flu'. A 'fresher' is the slang for a first-year student, but freshers' flu had no respect for seniority. It swept through the lot of us within the first couple of weeks of every term.

Eccles reasoned that at any given time, there must be plenty of students walking around with cold virus infections that they hadn't yet noticed. Some of those infections would be presymptomatic, meaning that they would develop into colds soon enough, while others would be asymptomatic, meaning that they would run their course without making the student ill.

If there was anything in the chilling hypothesis, then the difference between a presymptomatic and an asymptomatic infection might be a bucket of cold water.

He persuaded 90 students to sit with their feet in a bucket of 10°C (50°F) for 20 minutes, which was long enough to cause the closing off of the nasal blood vessels that Mudd and Grant had described nearly a century earlier. Eccles also recruited a control group who went through all the motions by sitting with their feet in an empty bucket for 20 minutes.

Freshers' flu, or whatever Cardiff students call it, must have been on the rampage because, within five days, 9 per cent of the students who got the empty bucket control had developed a cold. However, fully 29 per cent of those who had endured the cold feet treatment developed one[141].

It looked like Eccles had proved generations of British mothers correct: cold feet can cause a cold, albeit only in people recently infected with a cold virus. However, there was probably nothing particularly special about the feet. If chilling is important because it closes off those nasal blood vessels, then chilling any part of the body or the whole body is likely to be as cold-provoking as wet feet. Every one of my correspondents who had been warned to wrap up warmly had been getting reasonable advice and Marian Halcombe could indeed have been laid low by getting a good soaking while she was eavesdropping.

Eccles's result was more than a validation of generations of widely-held folk beliefs. It showed that the interactions between cold viruses and the immune system can be influenced by external factors, which leads us to the question of what other external factors may be involved.

We would not be the first to ask that question.

The invention of immunity debt

COVID-19 was, among many other things, the first major pandemic of the online era. With so many of us confined to our homes, the internet became our main source of information about a fast-moving situation. Some of that information was extremely valuable; many scientists and policymakers were exchanging information through websites and social media accounts that were open to anyone who cared to look at them. Unfortunately, some of the information being exchanged was somewhat less valuable.

In 2021, as 'lockdown' restrictions were eased around the world, the term 'immunity debt' sidled into the lexicon. It was often rolled out to explain the high levels of some viral and bacterial infections that followed the end of lockdown restrictions. 'When people avoided each other during the pandemic,' explained a *Wall Street Journal* article, 'they failed to build up the immunity against viruses that comes from normal contact'[142]. It read as though it was explaining a well-established immunological principle to a lay readership.

It was not.

That article was, in fact, attempting to define a neologism. Writing for the Office of Science and Society at Montreal's McGill University, one commentator observed that the concept of immunity debt 'sprang out of nowhere, but so many people act as if this is an old chestnut in the field of immunology'[143].

It's not entirely true to say that it came out of nowhere, although it seemed like it to immunologists being informed it was a cornerstone of their discipline. 'Immunity debt' first appeared in a paper published by a specialist group within the Société Française de Pédiatrie (SFP)* who were concerned that when lockdown restrictions were lifted, there would be a surge of childhood diseases[144]. As lockdowns had curtailed the circulation of SARS-CoV-2, they had also limited the circulation of many other infectious microbes. The authors warned that people who would normally have caught those diseases in the 2020–2021 winter had been protected by lockdown, making them likely to catch them in the 2021–2022 winter as well as all those people who were always going to fall ill at that time. Medical services, they warned, should prepare to deal with two winters' worth of several different diseases at once.

The authors used the term 'immunity debt', placed within inverted commas, as a shorthand for the phenomenon. They did not include a definition, presumably because they did not intend to coin a new term. In fact, they first used the term to state that immunity debt probably *wouldn't* cause an unusually high number of whooping cough cases.

It was a technical paper in a technical journal that would not normally have been noticed outside technical circles. However, those were not normal times. Someone at the *Wall Street Journal* evidently did notice it and liked the term enough to invent a definition. The article went on to quote Japanese and Australian doctors who were dealing with a very high level of respiratory syncytial virus, or RSV, a cold virus that often causes serious disease in babies and the elderly†.

As one country after another experienced higher than normal levels of RSV, influenza, scarlet fever and a host of other diseases, 'immunity debt' became invoked more and more frequently. Many of the people using it appear to have believed they were describing a well-established scientific principle rather than a term defined by a journalist in 2021 and, perhaps more importantly, few of them realised that the definition was not a very good one.

From the *Wall Street Journal* definition, it is not clear whether the journalist meant to describe a phenomenon operating at the level of the population or of the individual. In fairness, there had been some ambiguity in the SFP's paper.

---

* French Society of Paediatrics.
† Chapter 8 covers RSV in more detail.

Most of that paper described a population-level phenomenon. If someone caught a disease a year later than they would have done in the absence of lockdowns, it did not mean that individual would suffer a more serious bout of the disease than if there had been no COVID-19, no lockdowns and they'd caught it in the 2020–2021 winter. It simply meant that across the whole population, there would be more people in need of medical attention in the 2021–2022 winter.

However, the ambiguity of the definition led to it being used to describe a phenomenon affecting the individual. Without a constant stream of infections, the individual immunity debt argument went, the immune system falls out of practice. When that individual then encounters an infection, their unpractised immune system struggles to contain it and they get more ill than they would have done if regular infections had kept them in fighting immune trim.

Statements that the COVID-19 lockdowns brought on immunity debt at an individual level raised – or possibly begged – several questions. For one thing, most people in most countries experienced intermittent restrictions over the course of the year. There was plenty of opportunity for the less seasonal of the cold viruses, such as the adenoviruses and the rhinoviruses, to stay in circulation. Moreover, everyone's immune system is permanently occupied in containing the cornucopia of bacteria and long-term viruses that we all carry around with us. A few months without airborne infections won't leave our immune systems short of practice.

Another counterargument is that lockdown would only interrupt the circulation of infections with such brief infectious periods that they need large interconnected societies to exist. Such societies have only existed since the Neolithic Revolution between 10,000 and 15,000 years ago*. Proponents of individual immunity debt did not seem to realise that they were arguing our hunter-gatherer ancestors existed in a state of immunity debt for the first 285,000 of our species' 300,000 years until someone got round to inventing agriculture, enabling a society sufficiently prone to short-term infections to keep our immune systems in fighting trim.

---

* See Chapter 2 for a description of how the Neolithic Revolution enabled the existence of microbes with short periods of infectivity.

An article in the British newspaper *The Telegraph* made a rather desperate attempt to square the immunity debt circle by linking it to the high prevalence of colds in the winter, stating 'over the summer, the numbers … fall, leaving us again with an "immunity debt" in the winter'[145].

For that to make any sense, we would need to catch every one of the 200-plus cold viruses every winter to top up our immunity to each of them. Moreover, it assumes that cold virus infections are restricted to one season of the year when in fact, some cold viruses circulate continuously throughout the year in Europe and North America[146] and in some parts of China, there are not one but two influenza seasons in the year[147]. If there is indeed some annual cycle of immunity debt, a lot of viruses have missed the memo.

By the time *The Telegraph* published that article in December 2022, 'immunity debt' was at the centre of many a social media argument, much to the consternation of the SFP group that had inadvertently coined the term.

'Unfortunately,' they wrote in a follow-up article, 'due to possible misunderstanding and oversimplification of our concept, many controversies and polemics arose on social media and networks, and a backlash ensued.'[148]

It's tempting to dismiss the idea that immunity debt operates on an individual level but in fairness, the idea is not completely absurd. True, it assumes our hunter-gatherer ancestors would have lived in a permanent state of 'immunity debt', but they didn't have to contend with the short-term infections that the articles invoking immunity debt are referring to. Moreover, the SFP article pointed out that a few studies have shown that vaccination against or infection with one microbe can enhance the immune response against others[149], suggesting that immune stimulation may not be restricted to the microbe stimulating it.

Then again, given that the illness caused by cold viruses is caused by a misdirected immune response, any priming of the immune system would be at least as likely to push it toward the common cold constitution as away from it.

When one is descending into an ever-decreasing circle of inferences and counter-inferences, it's often helpful to ask whether anyone has asked the question that provoked them in the past and whether they did any investigation that went beyond inference. When the question relates to the common cold, the answer to that question is often, yes: Christopher Andrewes thought of it first.

Decades before anyone was arguing about immunity debt on social media, Andrewes was wondering whether a period of isolation from colds can make someone more susceptible to a cold when the protection is removed. His starting point was an early-twentieth-century report by a doctor based on the South Atlantic archipelago of Tristan da Cunha. The doctor reported that every time a ship arrived from Cape Town, the Tristanians would all be sniffling miserably within a few days, but ships arriving from anywhere else never brought colds with them[150]. The likely reason lies in the fact that if someone picked up a cold virus in Cape Town, it would pass around the passengers and crew for the week it took to get to Tristan da Cunha and someone could still be infectious when the ship docked. Ships arriving from anywhere else had been at sea for at least 12 days, which was long enough that everyone on board who caught the cold had got past the period in which it was infectious.

The isolation of Tristan da Cunha appeared to leave Tristanians unusually vulnerable to colds but then, the very fact that Tristan da Cunha was more than a week away from anywhere else made it an unusual situation. Tristanians were – and, given that there is still no airstrip on Tristan da Cunha, probably still are – protected from all the cold viruses that constantly circulate among the more connected majority of the human world, and so do not spend their early years building up the library of immunity to cold viruses that most of us carry around with us[151].

Which brings us back to the question of immunity debt. Does that immunity need to be boosted by continuous infections or does the immune system remember the cold viruses it's encountered before without needing constant reminding? It was a question that bothered Christopher Andrewes before he knew how many cold viruses the immune system would need to remember to protect against all of them.

In 1950, Andrewes's attempts to answer it would land CCU scientist James Lovelock in a very strange conversation.

Marooned students and the crofter's cold

In July 1950, a small boat chugged out of Skerray harbour, on the north coast of Scotland, and headed north beneath an unusually cloudless sky.

On board, the boatman asked Lovelock why they were taking a dozen students to an uninhabited island called Eilean nan Ròn.

Lovelock answered that it was to study the common cold.

The boatman chuckled. He wasn't the ignorant teuchter this well-spoken Sassenach took him for. He was carrying a dozen students with camping gear and enough food to spend the whole summer on Eilean nan Ròn and, barring emergencies, they were to be left alone out there. Whatever was going on, it had to be bigger than the common cold.

'C'mon', the boatman asked, 'what's it really all about?'

'It really is all about the common cold,' Lovelock insisted. 'Nothing secret about it.'

The more Lovelock insisted, the more he convinced the boatman it was a cover story. Eventually, Lovelock relented and told the boatman that the dozen students from Aberdeen University were searching for uranium.

That made sense to the boatman. Five years after uranium exploded into the public consciousness over Hiroshima and Nagasaki, the British government was bound to have people secretly scouring the Isles for the stuff.

He thanked Lovelock for being candid and promised to keep the secret mission under his hat, which convinced Lovelock that Scottish uranium would be the talk of every pub in the Highlands within the week[152].

Lovelock had been telling the truth in the first place; marooning the students on Eilean nan Ròn was Andrewes's way of investigating whether isolation made someone more susceptible to colds. Lovelock split the students into three camps of four people, instructed them to stay away from each other and left them to it. The experiment got off to a questionable start: one student brought a cold with him which he must have passed around, because four more of the students spent their first week coughing and sneezing. They all recovered soon enough and spent a healthy summer studying the island's wildlife.

Nine weeks later, the second phase of the experiment began. It was time to see if they could be infected with a cold. Andrewes and Lovelock took the boat out to Eilean nan Ròn to break the students' isolation. They spent some time with one of the camps to see if they had become so susceptible that they could catch a cold from a healthy person.

Four days later, no new colds had developed so they upped the ante, devising an approach that tested both contact and airborne transmission. Andrewes and Lovelock inoculated six volunteers with a cold and sent them to visit a second camp while the members of that camp were watching seals and seabirds. The six 'infectors' were, as Lovelock and Andrewes would report, 'liberal in the way they disseminated nasal discharge on playing cards, books, cutlery, handles of cups, letters, chairs ... and tables'[153]. If the four students had seen what was going on in their absence, they may have regretted signing up for it.

After three hours of 'discharging' all over the second camp, Lovelock took the six infectors to a sealed room in which they were separated from the members of the third camp by a blanket hung between them. The infectors coughed and sneezed enough to ensure the air was infused with anything infectious they breathed out and then spent the next three days with the same camp that Lovelock and Andrewes had visited.

It made no difference how the infectors tried to infect. No one in any of the three camps caught a cold.

With summer fading to autumn and the students needing to get back to Aberdeen for the beginning of term, Lovelock and Andrewes heard that a crofter living near Skerray had caught a cold. They knocked on his door and asked him to spend a day with one of the stubbornly healthy camps. At the same time, they brought another four people given colds at the CCU to spend a day with another camp. At the end of the day, the crofter and the new batch of infectors met at the third camp. The crofter spent a couple of hours with them before the boat took him back to Skerray while the infectors spent the next four days with them.

The crofter proved a more effective infector than any of the inoculated volunteers. Three of the four students who spent a full day with him developed colds. Once again, there was no sign of any of the students catching a cold from the volunteers who had been deliberately infected.

The Eilean nan Ròn experiment shared the characteristics of the soggy socks experiment: a solidly designed experiment undermined by doing it on too few volunteers.

Andrewes and Lovelock concluded that because the marooned students didn't catch their experimental cold, they had shown that isolation did not increase susceptibility – but what were they to make of the fact that some of the volunteers caught a cold from that crofter? Without having introduced him to an equivalent group of volunteers who had not been isolated on a small island, it was impossible to tell whether the isolation had increased their susceptibility, decreased it or made no difference.

Even with a comparison group, it would have been very difficult to compare groups of only four people. Suppose one or two of four un-isolated volunteers had caught the crofter's cold when compared to the three in the isolated group. A difference of one or two individuals would hardly be definitive proof that isolation increased susceptibility.

Such poor experiments were not uncommon in the mid-twentieth century, and it would be unfair to define either Andrewes or Lovelock by their poor grasp of sample size. Both made significant contributions to how we understand viral infection and indeed the world in general across very long careers. It is, however, worth reflecting on the weaknesses of the Eilean nan Ròn experiment because mid-twentieth-century cold research was bedevilled by such poorly constructed – or occasionally well-constructed but misinterpreted – experiments, and one still hears many a confident statement about colds that can be traced back to them.

The unbreachable lockdown

In 1976, Elliot Dick picked up the isolation question at the University of Wisconsin-Madison, as the University of Wisconsin had been renamed since he was orchestrating marathon poker games to study cold transmission*. By then, cold research was based on a much better understanding of the different cold viruses that had been gleaned since Lovelock set sail from Skerray and rather than recruiting a few student volunteers, Dick found a population of several hundred people who spent weeks under an unbreachable lockdown: the scientists and support staff who spent the winter in the American and New Zealand Antarctic research bases, where the weather cut them off throughout the winter.

---

* Chapter 3 describes the critical role of poker in elucidating how viruses are transmitted.

Antarctic veterans told Dick that when the weather cleared enough to allow new staff to be flown in, the first flight of the Antarctic spring was often followed by an outbreak of colds. Dick spent three years monitoring the bases and found that the spring arrivals did indeed bring outbreaks of adenoviruses and rhinoviruses, which are two of the most common causes of colds[154][155].

However, only a minority of the overwinter team caught the spring cold and even people who had no immunity to the recently introduced virus rarely became ill. It was hardly evidence of immunity debt building up in those weeks without exposure to cold viruses.

Moreover, Dick had found similar results when he monitored the postgraduate accommodation of his own university[156]. The University of Wisconsin-Madison's postgraduates hadn't been locked down for weeks but a rhinovirus outbreak revealed them to be, if anything, slightly more vulnerable to colds than the Antarctic overwinter crews. Evidently, the University of Wisconsin-Madison had its own version of freshers' flu.

Neither the Eilean nan Ròn experiment nor Dick's Antarctic observations provide any evidence that weeks of isolation make anyone any more susceptible to colds. One could argue that neither is a particularly convincing experiment. Moreover, both experiments had been based on adults but that infection-ridden winter of 2021–2022 hit children much harder, presumably because they had not built up a repertoire of immunity through infections before the advent of COVID-19.

Nevertheless, we can only evaluate the evidence we have, which does not support any suggestion that someone isolated from colds – or any other infection – will develop a more serious illness when they finally catch something. If anyone ever comes up with a proper definition of 'immunity debt', it will need to clarify that it operates at the level of the population and not the level of the individual. Then again, the term 'immunity debt' has become the source of so much confusion that it's probably best forgotten altogether.

The psychology of catching a cold

So far, we've established that getting chilled and soaked may well have contributed to Marian Halcombe's cold but even if she'd been isolated, it probably wouldn't have made any difference. There is a third factor to consider: she was under considerable psychological stress. If she was so worried about her dodgy brother-in-law and his scheming friend that she was sneaking around the roof in the pouring rain, perhaps she was worrying herself into a state that made her more susceptible to a debilitating cold.

The first evidence that stress might be a factor in the common cold constitution emerged out of the 'Asian influenza' pandemic of 1957. Six months before the pandemic hit, 480 employees of the US military's medical research establishment at Fort Detrick filled in a wide-ranging set of health and personality questionnaires. When the pandemic hit, employees whose answers got them classed as 'psychologically vulnerable' were three times more likely than their better-balanced peers to become seriously ill during the pandemic[157].

The pandemic arriving so soon after so many employees had completed such a survey was sheer good fortune – for medical knowledge, if not for those employees – and a pandemic influenza is not exactly a cold virus, although it would later evolve into one*. Nevertheless, the Fort Detrick survey strongly suggested a psychological element to respiratory infections like colds.

It was nearly two decades before anyone explored the subject any further. That exploration was begun by Richard Totman, a psychologist at Oxford University, who worked with the CCU in the 1970s. By then, CCU researchers had gained an understanding of the need for what statisticians call sample size which, in the CCU's case, meant the number of volunteers involved in an experiment. Totman's first experiment involved 52 volunteers who filled in questionnaires intended to assess their personalities and stress levels in their lives before being inoculated with rhinoviruses. Totman found that having either an introverted personality or relatively high levels of life stress made people more likely to develop a cold.

Stress and introversion made no difference to how likely a rhinovirus was to infect someone's cells. Totman and his colleagues found rhinoviruses replicating up the noses of as many stress-free extroverts as up those of stressed-out introverts. However, they found that both stress and

---

* Chapter 6 describes the mutation of severe pandemic influenza viruses into the cold-causing seasonal influenza viruses.

introversion made the infection more likely to cause a cold. Volunteers whose questionnaires revealed them to be extroverts or stress-free were far more likely to remain happily oblivious of their infection until their lymphocytes cleared it out[158].

After that, Totman moved away from cold research but where Totman left off, Sheldon Cohen of Pittsburgh's University of Carnegie Mellon picked up. Cohen built a transatlantic collaboration with the CCU to build on Totman's work. He evaluated a broader range of measures of the stress that volunteers were under in their day-to-day lives and he used a broader range of cold viruses in his experiments.

Cohen found the association between stress and colds was remarkably consistent. Regardless of how he measured stress or which cold virus he inoculated the volunteers with, volunteers under stress were more likely to develop a cold while those with relatively little stress were more likely to handle the infection without any symptoms[159].

Totman and Cohen had found a psychological element to the common cold constitution. Being under stress or simply having a certain personality makes one more likely to be oblivious to a cold virus replicating up the nose, and equally oblivious to the immune system kicking it out after a few days. While Totman had treated the basis of colds as part of a broader body of work describing the links between psychology and physical illness, Cohen saw his work with the CCU as merely a beginning. For him, it wasn't enough to know that being stressed makes a cold virus infection more likely to cause a cold. He wanted to know *how* stress affected the course of a viral infection. It became a major focus of his research for the next three decades.

His first problem was that the Medical Research Council axed the CCU's funding. After 44 years of infecting weekly tranches of volunteers, the CCU closed in 1990 and Harvard Hospital was demolished.

Cohen continued his research through a collaboration with Jack Gwaltney's research group at the University of Virginia in Charlottesville. Working with Gwaltney, Cohen found that infection was made more likely to cause a cold by smoking, poor sleep, exercising less than twice a week and, perhaps most interestingly, two hormones related to chronic stress called epinephrine and norepinephrine[160].

The involvement of hormones suggested a mechanistic connection between the volunteers' state of mind and state of body. Cohen explored that connection by exploring whether being worse off economically affected the likelihood of developing a cold after getting a cold virus up the nose. He measured how well off his volunteers were in two ways. First, he used the objective assessment of recording each volunteer's household income. Then he presented each volunteer with a picture of a nine-rung ladder and asked where they would place themselves compared to everyone else in the USA.

Cohen's first finding was that there was absolutely no correlation between his volunteers' incomes and where they placed themselves in comparison with everyone else. That's probably because most people had very little grasp of the overall economics of the USA. Each volunteer compared themselves to their social circles, which consisted of people in a similar income bracket to themselves.

One person might consider themselves as doing well if their 15-year-old car broke down less often than their neighbour's. Another might own two new cars and send their children to a private school but consider themselves deprived because their neighbour could afford all that and an annual cruise in the Caribbean.

When Cohen infected his volunteers with colds, he found that the likelihood of their developing a cold had nothing whatsoever to do with their income but it *did* correlate with which rung of the economic ladder they believed themselves to be on[161]. Feeling that they weren't doing well affected the volunteers much more than how well they were actually doing.

By separately assessing the volunteers' material circumstances and their perception of it, Cohen was able to conclude that it was stress itself that made them more susceptible to colds and not the things that were making them stressed. However, it still didn't answer the question of how, in physiological terms, stress pushed the volunteers toward the common cold constitution.

To dig deeper into that question, Cohen needed volunteers who were under prolonged extreme stress. He found them in the form of the parents whose children were being treated for cancer. They were under as much chronic stress as it was possible to be under and they spent a lot of time in the waiting rooms of the University of Pittsburgh Medical Center, which meant they didn't mind answering a few questions and baring their arms for a few blood samples.

Cohen didn't inoculate those parents with cold viruses. Cold viruses, cancer and chemotherapy make for a very dangerous combination and those parents had enough to deal with without having a cold. Instead, Cohen looked at how the immune cells in those blood samples responded to stimulation. One way that immune cells respond to stimulation is by producing cytokines, which are the molecules they use to communicate with each other. Cohen was particularly interested in the cytokines that encourage the sort of inflammation likely to lead to a cold.

At first glance, the cytokine profile of parents whose children had cancer looked identical to that of parents whose children were completely healthy. However, immune cell communication is an ongoing and complex conversation. At any given time, there will be cells secreting cytokines that both drive inflammation and inhibit it. The level of inflammation is a function of the balance between them.

Cohen and his colleagues applied a chemical that would normally mimic that suppression, and so it did for most of the inflammatory cytokines they measured except for one. The exception was a cytokine called Interleukin-6, or IL-6[162].

Now they had one cytokine to concentrate on, they moved their research out of the cancer ward and recruited another batch of healthy volunteers who they could inoculate with cold viruses. They found that volunteers who reported more stress in their lives were more likely to develop colds and were more likely to have immune cells that kept on pumping out IL-6 even when they were treated with an inhibitor[163].

It looked like Cohen had found the lever by which psychological stress flips a person into the common cold constitution. Further evidence would appear in 2020, when high levels of IL-6 production would emerge as a key difference between people who experienced COVID-19 as a particularly nasty but manageable cold and people who ended up in intensive care wards[164] or who suffered the long-term debilitation of long covid[165].

The infirmity of Marian Halcombe

If we consider Marian Halcombe's fictional cold in the light of more than a century of research since Mudd and Grant were putting corks between their volunteers' teeth, we can draw some conclusions about how she ended up so ill.

First, she would have needed a cold virus up her nose. After that, things could have gone either way. There was always a fair chance that the virus could have spent a few days replicating itself without bothering her in the slightest.

Also in Marian's favour was that she didn't smoke although, being a Victorian woman of the genteel classes, she probably didn't take much exercise except when she was clambering over roofs to eavesdrop on murderous conspirators.

She hadn't been more than usually isolated in the weeks before her soaking, but it wouldn't have made any difference if she had been.

As for whether getting rained on had anything to do with her cold, the jury is still out. Ronald Eccles's experiment in Cardiff strongly suggests that it did, but many other studies, going back to Andrewes's soggy socks experiment, concluded that it didn't make much difference. However, Eccles's experiment was by far the largest to consider colds acquired the natural way rather than squirted up a volunteer's nose. That gives it more weight than any other single study, if not quite enough to carry the argument on its own.

Another factor working against Marian was the psychological stress that drove her up to that roof in the first place. That sort of stress would have permeated every cell of her body, including those of her immune system, taking the brakes off their production of cold-inducing IL-6.

Seven decades before Alphonse Dochez showed that cold viruses cause colds, Wilkie Collins had created a surprisingly good case study of how someone ends up with a cold. Being a 'sensation novelist', he used the cold to flatten Marian at precisely the moment it would turn a bad situation into a disaster.

As for whether the ever-resourceful Marian would escape that disaster, there will be no spoilers here.

Chapter 5

Is the indoor environment an incubator for colds?

On 3 September 1939, Britain and France declared war on Germany. It was one of the most fateful days in modern European history. It was also the day that Owen Lidwell officially joined the Medical Research Council. He didn't present himself at the National Institute of Medical Research until the following day, not because of the anti-aircraft guns and barrage balloons being deployed in the middle of London, but because it happened to be a Sunday.* [166]

It marked a change of direction for Lidwell, who was a physical chemist by training, but he never looked back and spent the rest of his career in medical research. He stayed at the MRC throughout the war years and built a close friendship with James Lovelock, with whom he volunteered to be burned by flamethrowers to spare experimental rabbits[167]. He and Lovelock became the core of the MRC's Air Hygiene Unit and in 1946, they were dispatched to the CCU where he spent the next decade and a half working on cold transmission. While Lovelock focused on the mechanisms of transmission from one individual to another, Lidwell took a broader view to study the epidemiology of how colds move around a community.

He started in Bowerchalke, a Wiltshire village only a few miles from the CCU's base at Harnham Hill. During two years of meticulous observations, Lidwell found that school-age children were three times

---

* Lidwell's 1998 account of his joining the MRC was, 'I joined the MRC on the auspicious day of 1 September 1939. As it was a Sunday and the day war was declared, I didn't turn up until the following day'. However, the 1st was a Friday and it was the day that Germany invaded Poland, not the day that Britain declared war, so he'd probably confused the dates when he was talking about it nearly 60 years later.

more likely than adults to pick up a cold outside the household, while adults and infants below school age were more likely to pick up a cold at home.

Amid the thatched cottages of Bowerchalke, Lidwell's results revealed that schools were the major hub of cold infections. Children would pass their colds among each other at school and then take them home to infect parents, grandparents and any siblings who were too young for school[168].

It was a solid conclusion, but Lidwell didn't mistake it for a universal conclusion. Bowerchalke was one small village of around 200 people, most of whom worked in agriculture. Its small size made it relatively easy to track the movements of colds around the village, but as the 1940s gave way to the 1950s, life in such villages was looking more like the exception than the rule.

The urban incubator

Urbanisation is one of the great global trends of the last century.

Throughout the world, the last hundred years has seen people from rural areas packing their bags and moving to their countries' rapidly expanding cities. In 1950, as Lidwell was writing up the Bowerchalke study, slightly less than a third of the world's population lived in cities. By 2018, more than half the world's population lived in cities and by 2050, it's likely to be more than two-thirds[169].

Lidwell didn't need those projections to see how many rural Britons were leaving their villages and heading for the cities. If he wanted to know how colds moved around most of the world's population, Lidwell was going to have to study somewhere smoggier than Bowerchalke.

Lidwell turned his attention to several large offices in London and Newcastle. Each office hosted between 350 and 500 employees whom he pestered about their colds between 1951 and 1957, building a picture of how colds move around a far more urban community than Bowerchalke. He found that while urban schoolchildren exchanged cold viruses as enthusiastically as their rural counterparts, they didn't share the Bowerchalke children's near monopoly on infecting their parents. In London and Newcastle, adults were as likely to introduce a cold into a household as schoolchildren[170].

In Chapter 2, we saw how cold viruses need large and interconnected communities to sustain them. Lidwell's research showed that by making societies even larger and more interconnected, urbanisation changes the way that colds spread around a community.

A cloud of virus-laden respiratory droplets disperses very quickly outdoors shut into a room, it can hang around for a very long time. School classrooms shut clouds of respiratory droplets into rooms full of children who are still building their library of immunity to cold viruses. No wonder Lidwell found that in rural Bowerchalke and urban London and Newcastle alike, children were sharing their colds at school and taking them home to infect their families.

The difference was that those urban hubs offered far more opportunities for cold viruses to pass from one adult to another.

The residents of Bowerchalke might get together in the pub, the church or the bus to Salisbury, but they spent far less time sharing indoor spaces than people who spent their nine-to-five in an office building with dozens or hundreds of other people. Moreover, many of those office workers got to those offices on buses and trains that were packed full of fellow commuters, all of whom spent the rush hour breathing each other's respiratory droplets.

Lidwell showed that as people flock to the world's cities, they are flocking into an environment that is perfect for the transmission of colds, or indeed any other respiratory microbe.

In the years that Lidwell spent pestering those office workers, he and Lovelock were among the few scientists working on respiratory virus transmission. In the USA, William Wells was preoccupied with caring for his son, and few other scientists were interested. That changed in 1957, the year in which Lidwell completed his office study, when an influenza pandemic killed between one and four million people[171]. It was nowhere near as devastating as the 1918 influenza pandemic, or indeed the COVID-19 pandemic that would arrive six decades later, but it was enough to shove respiratory virus transmission up the agenda. In 1968, another influenza pandemic underlined the point by killing a similar number of people[172].

Lidwell had left the CCU by then, but his work showed how urbanisation was making the world more vulnerable while pandemics remained both inevitable and unpredictable. Unfortunately, policymakers had not taken that on board until those pandemics drove the issue up the research agenda although, even today, persuading policymakers to consider pandemics unless they're in the midst of one remains an uphill struggle.

The winter cold season

While Lidwell was tracking colds through the schools and homes of Bowerchalke and Newcastle, he found that people of all ages were more likely to catch a cold during the winter than in the summer. That didn't surprise him; coughing and sneezing was as much a British winter tradition in the 1950s as it is now. At the time, Lidwell didn't know that it's not a tradition exclusive to Britain. Colds are more prevalent in the winter across the mid and high latitudes of the entire northern hemisphere. The winter cold season that Lidwell recorded in Newcastle was just as regular in New Mexico and Nagasaki[173].

Anyone who lives north of the tropics has noticed that there are more colds between October and April but what's not so obvious is that, as the figure on page 76 shows, different viruses follow different seasonal patterns. The winter cold season occurs because more viruses peak in the winter than at any other time rather than because winter favours all cold viruses indiscriminately. The parainfluenza viruses and the rhinoviruses, for example, are more prevalent in the spring and the autumn than in the coldest months. Summer colds are most likely to be caused by the adenoviruses and the bocaviruses while COVID-19 has never shown any consistent seasonal pattern.

While the northern hemisphere is enjoying a respite, colds make their way to the southern hemisphere's temperate zone, sending colds sniffling their way through New Zealand and the southern parts of Australia, Chile and Argentina. It's not possible to draw a corresponding figure of what the different viruses do in the southern hemisphere because there's far less data to base it on, but it's likely that each virus follows much the same pattern in the south as it does in the north, albeit offset by six months.

So far, so predictable. Cold weather brings colds. It's a clear answer to the question of when to expect colds in temperate latitudes – or rather, it looks clear until we look between the temperate zones to see what's happening in the tropics.

	Jun	Jul	Aug	Sep	Oct	Nov	Dec	Jan	Feb	Mar	Apr	May
influenza							●	●	●			
endemic coronaviruses							●	●	●	●		
RSV						●	●	●	●			
adenovirus & bocaviruses	●	●	●	●	●	●	●	●	●	●	●	●
parainfluenza	●			●	●	●				●	●	●
metapneumovirus				●	●	●	●	●	●	●		
rhinovirus				●	●	●	●	●	●	●	●	
COVID-19	●	●	●	●	●	●	●	●	●	●	●	●

Seasonal prevalence patterns of different viruses in the northern hemisphere temperate zone. Far more viruses are prevalent in winter than at any other time although there are exceptions: the parainfluenza viruses and rhinoviruses are more prevalent in the spring and autumn, the adenoviruses and bocaviruses are present throughout the year and COVID-19 prevalence varies unpredictably without following any consistent seasonal pattern.

Wherever colds have been studied in the tropics, prevalence has been found to follow a seasonal pattern. However, different seasonal patterns have been found in different regions, even when those regions are close together. It doesn't help that far less cold research – or medical research in general – is done in the tropics than in the temperate zones, but there are more than enough results to get confused by.

One of the better-studied cold viruses is the respiratory syncytial virus, largely because it can cause illnesses far more dangerous than a cold in young children*. Consequently, we have enough data to know

---

* Chapter 8 describes the respiratory syncytial virus in detail.

that it follows very different seasonal patterns in different parts of Africa, with at least as much variation between tropical Asia and the tropical Americas[174]. So far, no one has found any factor or combination of factors that consistently drives those patterns.

In 2013, a team led by Hongjie Yu of the Chinese Centre for Disease Control and Prevention dug into the seasonal patterns of influenza, another of the more thoroughly studied cold viruses, across China[175]. Across the temperate northern regions, Yu found what he would have expected: influenza followed the typical temperate pattern in which most cases occur in the winter months of January and February. However, the tropical southern regions showed a very different pattern with most cases happening in May and June. In the sub-tropical regions across central China, Yu found not one but two peaks of influenza prevalence. One coincided with the temperate peak in January and February while the other coincided with nothing much at all, between June and August.

Yu's data refers to one type of cold virus in one country, albeit one of the largest and most climatically diverse countries in the world. Nevertheless, it defies any simple explanation for *why* influenza follows seasonal patterns that are simultaneously so variable and so predictable.

Returning to the temperate zones where most studies of cold epidemiology have been done, we're left with the question of *why* colds are more prevalent in the winter. True, Ronald Eccles showed that chilling may well be a factor when he persuaded students to put their feet in buckets of cold water* and true, a denizen of the temperate zones is more likely to get chilled in the winter. However, Lidwell's office workers were not putting their feet in buckets of cold water. They were far more likely to avoid being chilled by heating their homes and offices to a comfortable temperature and donning coats and gloves when they went outside.

Nor were they likely to be spending much more time indoors. During the summer, they might have opted to have the occasional drink in their local pub's beer garden instead of the snug, and some probably cycled to work instead of taking the bus. Nevertheless, they still spent

---

* Ronald Eccles's cold feet experiment is described in Chapter 4.

every working day in their offices and their children spent every school day in their classrooms.

Urban living has seen some changes since the 1950s. More people drive to work, avoiding the mist of respiratory droplets that suffuse public transport during the rush hour, and more buildings are heated, placing the indoor environment under the control of the people who live in it rather than the whim of the seasons – up to a point.

Heating makes the temperature inside our buildings a lot more comfortable in the winter, but it doesn't divorce the indoor environment from the outdoor. From the perspective of virus transmission, there is one key aspect of the indoor environment that is very dependent on the outdoor environment: humidity.

Heating and the cold virus incubator

There are two ways to measure humidity: absolute and relative. Absolute humidity is a simple measure of how much water vapour is in the air. It's as simple as it sounds but, as you've probably noticed, very little about catching colds is that simple. You won't be surprised to know that it's the more complicated measure of relative humidity that matters.

Relative humidity is a measure of how much water vapour is in the air compared to how much water vapour the air can hold. The warmer the air, the more water vapour it can carry.

To understand the difference between absolute and relative humidity, imagine it's mid-December and we're following the air from the streets of Newcastle into one of the offices that Lidwell was studying. The office's heating raises the temperature of the air without affecting how much water vapour is in that air. However, heating the air has increased how much water vapour the air *could* hold so, although the absolute humidity has not changed, the relative humidity has dropped.

The drop matters because, at a comfortable room temperature, cold viruses transmit best at low relative humidities[176]. That's partly because the lower the relative humidity, the faster the evaporation rate. It's why laundry dries faster in the high temperatures of an airing cupboard and, as William Wells showed in the 1930s*, the faster water evaporates off

---

* William Wells's pioneering experiments on the physics of airborne droplets are described in Chapter 3.

the surface of a droplet, the more likely that droplet is to shrink to a size where it can stay airborne before it hits the ground.

Let's return to that imaginary office and consider the people working in it. Perhaps there's a typist who picked up a cold virus on the trolleybus she took to work yesterday. Now she's going about her day's work, blissfully unaware of the raging sore throat she'll wake up with tomorrow and of the cold virus suffusing the droplets she's breathing into her office.

What happens to a droplet after she's breathed it out depends on how large it is. A tiny droplet of five micrometres across will be small enough to float around the office for a couple of hours. A large droplet of 200 to 300 micrometres will fall straight to the ground. Any virus it contains is unlikely to bother anyone for the hours or, at most, days that it will last down there.

The typist's droplet may hit something on the way, like her typewriter or the report she's typing on it. From there, it may find its way onto someone else's hands and, if they touch their face, into their mouth or nose. We can only hope that whoever she's about to give that report to doesn't smoke while they read.

Relative humidity is unlikely to make much difference to the largest or the smallest droplets she exhales, but it can make a substantial difference to medium-sized droplets. To see how, let's focus on one droplet that's 90 micrometres across when it comes out of her nose, which places it squarely in the mid-range of respiratory droplet sizes. It's large enough that it will fall toward the ground but small enough to take its time about it.

For the sake of simplicity, let's say that our typist happens to exhale it while she's standing on her chair to get a file off a shelf. That places her nose and mouth at the conveniently round number of two metres above the floor although, as the metric system had yet to arrive in British offices, she'd call it six feet and three inches. From up there, that 90-micrometre droplet will take about nine seconds to hit the floor[177] – but only if it remains a 90-micrometre droplet.

If the air were close to carrying capacity, at perhaps 80 or 90 per cent of relative humidity, not much water will evaporate off it and it won't shrink appreciably in nine seconds. Unless our office worker breathes it directly into the face of a colleague, the droplet will carry the virus down to the well-worn carpet where it won't bother anyone.

However, the air in Britain very rarely reaches 80 to 90 per cent of relative humidity. It's usually between 40 and 60 per cent, which happens to be the level that most people find most comfortable[178]. In the summer, the office heating will be switched off so the relative humidity indoors will be similar to the relative humidity outdoors. As the figure on page 81 shows, that's low enough that the droplet starts evaporating as soon as our typist exhales it. As it shrinks, the droplet will fall more slowly, only hitting the ground after around 15 seconds. It's appreciably longer than nine seconds but it's still not very long. The chances are that it will hit the ground before anyone inhales it.

In the winter, however, that droplet will follow a very different trajectory. Although the outdoor relative humidity is much the same as it would be in the summer, the heating in our typist's office has driven the air temperature up and the relative humidity down to around 30 per cent. Now the water is evaporating off that droplet so fast that before it hits the ground, it shrinks to a size at which it stops falling. It's now become one of those tiny droplets that can float around the office for minutes or hours at a time, floating its cargo of cold viruses around the office.

It's an identical droplet exhaled from an identical height but, because the heating has driven down the relative humidity, it has vastly more time to find its way up someone's nose than its summer counterpart that hit the carpet in 15 seconds.

Drying the mucociliary escalator

Low relative humidity is good for cold virus transmission in another way: when one of the typist's colleagues breathes in her virus-laden droplets, they will be more prone to being infected. That's because low relative humidity disrupts the flow of mucus through the respiratory tract.

The entire respiratory tract, from the bronchioles up to the mouth and nose, is coated with mucus. We tend to regard it as a nuisance because we only notice it when a condition like a cold or hay fever makes us produce too much of it. Most of the time, it's quietly protecting us by trapping all the particles carried by the roughly six litres of air we breathe every minute just to stay alive. Any sort of activity pushes our breathing up to around 16 litres of air per minute[179], and all that air comes laden with dust, pollen grains and floating droplets full of cold virus.

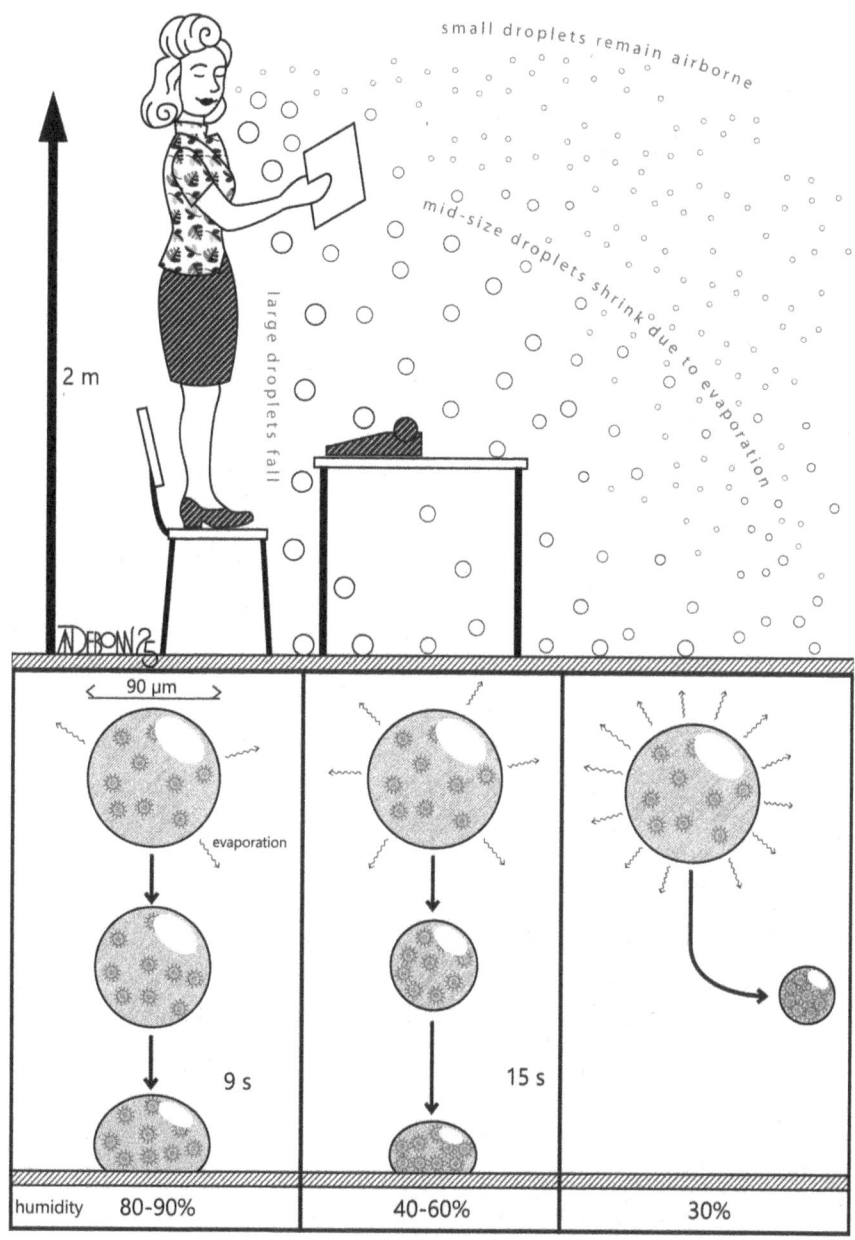

In an office of the 1950s, a typist is exhaling droplets containing cold viruses. The lower the relative humidity, the more of those droplets will evaporate down to a size at which they can remain airborne before they hit the ground.

The mucus of the respiratory tract is constantly on the move. The cells that make up the inside layer of the respiratory tract have hairlike structures called cilia, which wave back and forth in a coordinated manner to waft the mucus upward into our mouths, from where we swallow it into the virus-dissolving acid of the stomach. The mucociliary escalator, to use the rather unappetising technical term, protects the respiratory tract cells from airborne microbes of all kinds but it doesn't function well when relative humidity falls below 40 per cent. A dry mucociliary escalator is in fact one of the main reasons why low relative humidity is uncomfortable.

Above 40 per cent relative humidity, the mucus is deep enough that a virus caught on its surface is unlikely to find its way down to the cells. That's just as well; it's those ciliated cells that most cold viruses infect. Moreover, the mucus is suffused with antibodies against cold viruses that we've been infected with before. Those antibodies may destroy the virus outright or may make it more difficult for it to infect a cell which, combined with a healthy mucociliary escalator, can be the difference between having a cold and not being infected at all.

The relative humidity matters because mucus is mostly water, thickened into the gooey substance we recognise by a concoction of biomolecules[180] – as long as the relative humidity is comfortably above 40 per cent.

The figure on page 83 shows what happens when the relative humidity drops below 40 per cent: water evaporates from the mucus faster than it is replaced. The gooey mucus layer becomes less gooey until it becomes so dense that the cilia struggle to keep it moving. At the same time, the mucus becomes shallower, drawing trapped viruses closer to the cells that healthy fluid mucus protects.

By increasing the evaporation rate, low relative humidity gives cold viruses a helping hand in two different ways. To return to the infectious typist standing on the chair, the low relative humidity keeps far more of her infectious droplets airborne for longer, making it more likely that one of her colleagues will inhale them. At the same time, the low relative humidity has rendered that colleague's mucociliary escalator shallow and sluggish and much less able to keep the cold virus away from the cells it needs to infect.

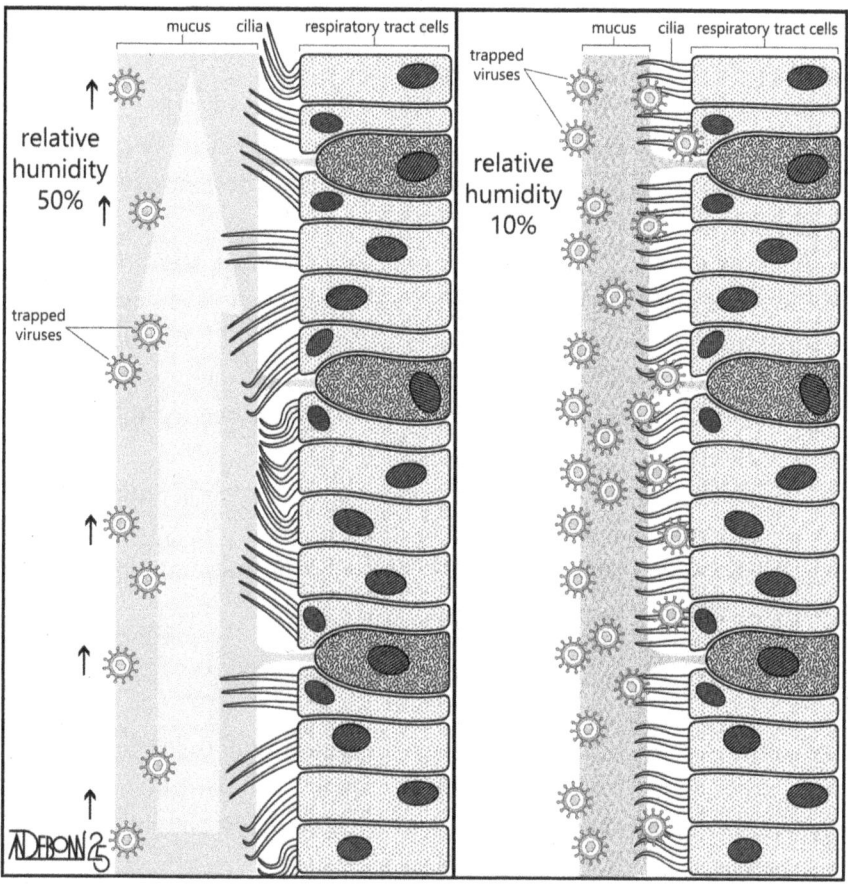

The effect of relative humidity on the mucociliary escalator. At a comfortable 50 per cent, the mucus layer traps inhaled viruses and the rhythmic pulsing of the cilia keeps it flowing up to the mouth, from where it is swallowed into the stomach. At an uncomfortably dry 10 per cent, evaporation leaves the mucus layer shallower and denser, drawing viruses closer to the cells they infect and immobilising the cilia.

Low relative humidity is a problem that suggests a solution: raise the relative humidity into the 40–60 per cent comfort zone. In the winter of 1969–1970, with the 1968 influenza pandemic still at the forefront of every public health professional's mind, one Charles Sale gave it a try by placing humidifiers in Virginia nursery schools. He found that humidification cut absences by around a third[181]. He didn't record what was keeping children at home but as colds are one of the most likely causes of school absences[182] and one of the most likely causes to be prevented by humidification, Sale's humidifiers probably prevented a lot of colds.

Nobody followed up Sales's success. There's no obvious reason why not, other than a tendency for indoor transmission to be prioritised immediately after a pandemic and to fall down the priority list as the memory of each pandemic fades.

Anecdotally, I've found that talk of humidification can alarm building engineers. When I've broached the subject, I've been frowned at and told that raising humidity encourages mould. Pointing out that mould rarely grows at relative humidities below an uncomfortably stuffy 80 per cent[183], far higher than the 40–60 per cent needed to limit cold viruses, rarely mollifies them.

Unfortunately, discussions of preventing indoor transmission rarely get to that stage. The idea of humidifying public indoor spaces has not caught on and cold-virus-friendly low humidities remain the norm throughout the winter.

The economics of catching a cold

Our cities are the engines of modern economic activity. For each of us as individuals, they are where we're most likely to find a job and a salary. Even businesses that operate in rural areas are largely dependent on customers, or possibly their customers' customers, who are concentrated in cities. For a country as a whole, cities are the economic hubs that generate taxes that pay for roads, schools, hospitals and everything else that makes up the infrastructure of a modern country.

Economic activity does not simply happen. It is the product of people working, trading, buying and doing all the other things that drive the nebulous entity that is 'the economy'. People tend to be better at doing things, including those things that economists call activity, when they are healthy. Yet if someone were to sit down and plan a way of living that was ideal for transmitting colds, they would come up with something that looks remarkably like a modern city. On a typical day, a modern city-dweller may take a bus or a train to a workplace they share with several hundred people, pass through a supermarket checkout queue on their way home and then go out to a pub or cinema. By the end of the day, they've breathed in the respiratory droplets of dozens or possibly hundreds of other people. If any of those people were exhaling a cold virus, that virus is now up their nose. They'd better hope their mucociliary escalator is in good shape which, during the winter, it probably isn't.

By turning our cities into hubs of both economic activity and cold transmission, humanity has created an irony that could break an economist's heart: colds are a significant drain on economic activity.

A cold can impose an economic cost through days off work, either to recover from a cold or to care for a child who is too ill to go to school or daycare. The scale of that cost is difficult to calculate, although a few intrepid spreadsheet-wranglers have given it a try. In 2003, a team from the University of Michigan totted up what colds cost the American economy through workplace sickness, caring for sick children and healthcare costs. They calculated that it came in somewhere between $31 billion and $48 billion[184].

However, the Michigan team excluded the effect of influenza from their dataset. Their rationale was that influenza was distinct from the other cold viruses because it causes more illness and incurs more cost. It's a somewhat questionable rationale because seasonal influenza viruses are transmitted in the same way as any other cold virus and cause an illness that's impossible to distinguish without diagnosing the virus causing it. The main consequence of that decision is that their $31–48 billion estimate captures only some of the cost of colds.

A few years later, a Communicable Disease Center team approached the same question from the opposite end; they calculated the economic impact of the influenza virus while excluding the other cold viruses. Influenza proved so much nastier than the other cold viruses that the CDC's estimated cost of between $47 and $150 billion[185] places the lower end of the estimate for the cost of influenza only slightly lower than the higher end of the Michigan team's estimate of the cost of all other cold viruses put together.

Not being an economist, price tags in the billions tend to make my eyes glaze over. I need to understand them in terms of what they mean for a typical person who has to earn a living, pay taxes, buy groceries and generally do typical-person things. For that typical person, there isn't a meaningful difference between a cold caused by influenza and any other cold virus, so I'd really like to start with an assessment of all cold viruses. In the absence of such an assessment, the best I can do is reach for an envelope and scribble some calculations on the back of it.

When I add the estimates for influenza and non-influenzal colds together and adjust for inflation, my back-of-the envelope calculations tell me that in 2025, colds cost the average American between $475 and $1,200*.

If that isn't enough to wash the glaze from the most economically indifferent eyes, it's likely to be a considerable underestimate of the impact of colds today. Both of those estimates were calculated before the emergence of COVID-19 which, like influenza, is effectively a severe type of cold.

In 2023, a London-based consultancy called the Office of Health Economics (OHE) calculated the first and, at the time of writing, only estimate of the effect of all types of colds in a world of work that harbours – or, all too often, nurtures – COVID-19. The OHE team estimated that the typical British worker is affected by a cold for around five days every year. That average worker takes one day off and spends the other four as a 'presentee', turning up for work that they don't do as well as when they're healthy.

For a mid-sized business with 100 employees, the OHE estimated that colds would cost an eye-watering £85,000. When they extrapolated to the entire British economy, they calculated that colds drain away £44 billion† every year[190]. At the time the report was published, that was nearly a quarter of what the British taxpayer spends on health and social care and more than the total national spend on defence[191].

Working out the cost to the typical Briton is more difficult because, while the Michigan and CDC estimates aimed to capture the cost to the whole economy, the OHE assessed the impact on the much narrower outcome of workforce productivity. The Michigan and CDC estimates included the cost of healthcare, which the OHE did not, and the OHE only included the impact on people outside the workforce if it impacted people within it. Moreover, the OHE estimate did not include any

---

* The University of Michigan study drew data from between 2000 and 2001 while the CDC study used 2003 data. The quoted figures are calculated by adding the costs calculated by the two studies, dividing by the 2003 US population[186] and inflation adjusting from 2003 to 2025[187]. In 2025, $475-1,200 was equivalent to £346-874 or €403-1019[188].

† In 2023, £85,000 was equivalent to $100,000 or €99,000, and £44 billion was equivalent to $56 billion or €51 billion[189].

healthcare costs and did not consider people who were too young, too old or too disabled to be employed.

To illustrate the difference, imagine the very common situation of a parent taking time off work to care for a cold-afflicted child, and then imagine that parent is worried enough that they take the child to see a doctor. The Michigan and CDC estimates included the cost of the parent's time off work to themselves and their employer, the cost of the doctor's appointment to the healthcare system and the cost of medications the parent would probably buy from the pharmacy. The OHE estimate only captured the cost of the parent's time off work.

That also means that the OHE estimate did not consider people like Lucy, who we met in the introduction, who was so severely debilitated by influenza and COVID-19 that she was forced to leave the workforce altogether.

With those caveats, the OHE estimate for lost productivity works out at £642* for every British worker, which is roughly in the middle of the total economic burden of colds in the pre-COVID-19 USA even though the OHE calculation only captured some of the costs captured by the Michigan and CDC estimates.

From the scant information that we have, it looks likely that if the CDC or Michigan calculations were applied today, the cost would be well above that $475–$1,200 per person derived from pre-COVID-19 data.

That's a conclusion delivered with a great deal of handwaving. Even if I hadn't needed the back of an envelope to pull it all together, direct comparisons between different countries can only ever go so far. There are many differences between the way the British and American economies are structured that colds may impact in a different way.

If you're an economist, you will have been wincing your way through those last few paragraphs, for which I apologise – but my apologies will be less than profuse. I would like to know why economists have been so uninterested in a multibillion-dollar ball and chain that I need the back of an envelope in the first place. I would love to be able to quote up-to-date estimates of the cost of colds to the residents

---

* The OHE calculation was based on 2023 data, when £850 was equivalent to $1,078 or €988.[192]

of different countries, but those estimates are simply not there to be quoted.

And those estimates matter because those billions of dollars are the sum of the hundreds – or, more likely, thousands – of dollars that colds cost every one of us. Lurking behind those billions of dollars – or pounds or euros or kwacha or whatever currency you happen to use wherever you happen to be – are millions of human stories.

There's the story of the zero-hours contract worker who loses a day's pay they can't afford because they're too ill to go to work or because they have to look after a cold-ridden child.

The story of a small business that's struggling to get by at the best of times and struggles even more when the winter influenza outbreak keeps two of its three employees at home.

The story of struggling through a job interview or critical meeting and never knowing whether the cold that wouldn't let you think straight was the reason it ended with, 'don't call us, we'll call you'.

The story of the child whose asthma is manageable except when they get a cold, which happens so often that they're struggling to follow the classes they're now attending.

The story of dragging a cold into work despite knowing that your work will be substandard and you'll probably infect your colleagues, making you both a contributor and a victim to the so-called working-while-sick epidemic[193] that particularly afflicts Britain, where such 'presenteeism' is far more of a workplace norm than in most other European countries[194].

They're such common stories that, although I've never met you, I'm sure that you or someone you care about has lived at least one of those stories. I'm also sure you have a few stories of your own that you could add, in which a cold has impacted a life to a far greater degree than the few days of sniffles that probably come to mind at the first mention of a cold.

The good news is that there are a few things we can do to treat or avoid colds, which will be discussed in detail in the third section of the book. Before we get to that, it's time to meet our adversaries through the century of research that has revealed them.

Section 2

The Discovery of the Cold Viruses

Chapter 6

The discovery of influenza

In 1950, Johan Hultin and his class of medical students trooped into a University of Iowa lecture theatre to listen to a guest speaker. Hultin hoped to learn from the speaker's expertise on influenza. He didn't expect to hear something that would send him to the edge of the Arctic Circle not once but twice.

The guest spoke of the mystery surrounding the 1918 pandemic, when an influenza virus had killed tens of millions of people[195], and then vanished. The key to that mystery, the guest believed, was to get a sample of the influenza virus that caused the pandemic. He could think of only one way of doing so: 'Someone ought to go to the frozen north to find a victim from 1918 in a permafrost grave.'

The guest meant it as an off-the-cuff comment. He was speaking to an audience of doctors and laboratory scientists, not polar explorers. Nevertheless, his words made Hultin sit up a little straighter.

Hultin was not a typical midwestern medical student. His classmates realised that when they were trudging to their classes through the snow and ice of their first freshman winter and Hultin glided past them on a pair of cross-country skis. It made him an unusual sight in Iowa but Hultin had grown up in northern Sweden, where going to school in winter had involved skiing four miles each way[196].

Before starting his degree, he and his wife had travelled through every state in the USA and, when Hultin thought back to their time in Alaska, the guest lecturer's idea seemed a lot more achievable than the guest himself seemed to realise.

The following year, Hultin was scouring Alaska for the right permafrost grave. The 1918 pandemic had hit some communities harder than others and none more so than the Alaskan Inuit. The territorial government had resorted to hiring gold miners, who knew how to excavate the rock-hard permafrost, to travel from settlement to settlement and inter the dead in mass graves[197].

Hultin found one such grave in Brevig Mission, on the edge of the Bering Strait, where the 1918 pandemic killed 72 of the 80 residents[198]. The survivors abandoned Brevig Mission, leaving a single wooden cross to mark the last resting place of the people who had once lived there.

It was exactly what Hultin was looking for. He arrived in the nearby town of Teller by floatplane and explained his quest to the village council. Three decades on, memories of 1918 were still raw. The council was willing to help a man whose research might help, in some small way, to stop it from happening again.

They ferried Hultin to Brevig Mission in a walrus-skin whaleboat and helped him to exhume the bodies. It was no easy task. They had to melt the permafrost with driftwood fires, dig until they could no longer get their shovels into the ground, then light another fire and start again. On the fourth day, they found the frozen remains of a girl with ribbons in her hair. She was the first of four victims that Hultin and the volunteers managed to exhume. He collected samples from their lungs and returned them to their shared grave.

Now it was a race against time. Hultin needed to get the samples back to Iowa City before the ice in his thermal jugs melted. Crossing half of North America by propeller-powered airliner took a worrying amount of time. More than once, Hultin had to purloin carbon dioxide snow from airport fire extinguishers to keep his samples frozen[199].

Hultin managed to keep the samples frozen all the way to the university but then he ran into the limitations of 1950s virology. There were only two things a virologist could do with influenza: infect a ferret or infect a fertilised chicken egg. Nothing in any of Hultin's samples would do either. If there had ever been any influenza in the lungs he'd dissected, it was no longer active.

Hultin's first quest for the 1918 virus had been a failure. It wasn't his last, although he wouldn't know that for another 40 years.

The seasonal influenza

If you asked a cold expert which cold virus was the nastiest before 2020, they would all give the same answer: seasonal influenza. Since then, it's lost the top spot to COVID-19, although seasonal influenza remains as nasty as it ever was.

One American doctor reportedly diagnosed influenza by asking, 'Do you feel like you've been hit by a train?'

If the patient said no, they didn't have influenza.

If they said yes, he'd ask, 'Have you been hit by a train?'

If the answer was no, he'd diagnose influenza[200].

The possibly apocryphal story shows how 'flu' has become a colloquial term for a bad cold, which is somewhat misleading. Most influenza infections cause colds that are indistinguishable from infections with other cold viruses and many cause no symptoms at all[201], while other cold viruses can cause infections that make you feel like you've been hit by a train.

However, influenza does damage the respiratory tract more than most other cold viruses. When you get a sore throat from a rhinovirus[202], it's because it's triggering an inflammatory response back there. When an influenza virus gives you a sore throat, it triggers the same inflammatory response but it also kills the cells in which it replicates[203]. Those include the ciliated cells that keep the mucociliary escalator* moving, without which the mucus builds up until it's shoved upstairs with a hearty cough[204]. After a week or two, the coughing feels less hearty than downright exhausting – and the coughing isn't the biggest problem that a damaged upper respiratory tract can cause.

Like all of our internal and external surfaces, our respiratory tract is home to a thriving community of bacteria. However, bacteria that are benign tenants on the respiratory tract surface can be a lot less benign if influenza strips away that surface and lets them underneath it. That's why influenza infections are so often the gateway to painful and debilitating ear and throat infections, especially in young children[205]. Worse, there's a risk that they may spread down to the lungs and cause a life-threatening pneumonia[206].

Seasonal influenza is a nasty piece of work. There can't be many people whom it hasn't sent at least as far as their GP, usually with one of those secondary bacterial infections, and it often fills hospitals during its seasonal outbreaks.

---

* Chapter 5 has a more complete description of how those cilia keep the mucus moving up the respiratory tract and into the mouth so it can be swallowed down to the stomach.

It's called seasonal because in any given part of the world, it causes an outbreak at roughly the same time every year. In the high and mid-latitudes, it's one of the many cold viruses that arrive during the winter but it spreads much faster than those other cold viruses. For eleven months, there is very little influenza but when it arrives, it blazes through any location within four to six weeks.

In that time, it makes between one in five and one in ten people ill[207]. Most of its victims will recover after a few days of curling up with some hot honey and lemon. However, one in every few hundred will need hospital treatment and because the outbreaks are so 'explosive'[208], all of those unfortunates need a hospital bed at the same time. That's the same time that between one in five and one in ten doctors and nurses are curled up at home with their honey and lemon or, in some cases, taking time off work to care for an influenza-ridden child.

Seasonal influenza's explosive spread fills hospitals while rendering them understaffed which is why, in many countries, it also generates dire headlines of an annual crisis in the healthcare system.

If that was the worst influenza can do, Hultin would not have gone to Brevig Mission. However, Hultin was very aware that seasonal influenza is only a taste of the devastation that influenza viruses can unleash.

The Spanish Lady

When the USA joined the First World War in 1917, it needed a bigger army and it needed it fast. The War Department's solution was to throw up massive training camps. From sea to shining sea, young men shambled through makeshift gates as civilians and marched out a few weeks later as soldiers. Few were sorry to leave. The camps where the 'doughboys' learned to march and shoot were overcrowded and had only the most basic amenities.

On 4 March 1918, a recruit reported sick in Camp Funston, which sprawled across the Kansas prairie. Illness in a training camp was nothing unusual; the constant influx of men ensured a constant influx of infections, and packing men together in barrack huts ensured those infections spread rapidly. Neither that recruit nor the medical orderly who examined him could know they were dealing with the first recorded case of the first influenza pandemic of the twentieth century.

There were between 50,000 and 60,000 men in Camp Funston and within three weeks, around 1,100 of them were in the camp hospital.

Thousands more were treated in improvised infirmaries or were left to sweat out the fevers and racking coughs in their bunks[209].

By then, the influenza virus that ailed them was spreading around the world as swiftly as the annual seasonal influenza outbreaks that we're familiar with today, reducing populations to coughing, sniffling wrecks for a few weeks and then passing on. By the summer of 1918, most of its victims had recovered and returned to what they had been doing before, be it farming, manufacture or fighting the First World War.

Not many died that spring, but the pandemic was only getting started.

In August, a new wave began an equally rapid but far more deadly spread. The 'autumn wave' lasted until December and accounted for most of the fatalities of the 1918 influenza. The pandemic still hadn't finished with the world; Spring 1919 saw a third wave, after which influenza subsided back into its seasonal pattern.

Overall, the 1918 pandemic influenza killed around one in 40 of the people it infected. Given the vagaries of national record keeping at the time, it's impossible to know the total death toll, but it's estimated at between 50 and 100 million. The world's population was around a quarter of what it is now, which meant that it was a very lucky survivor who was not left mourning a close friend or relative[210].

In the years after the pandemic, it became known as the Spanish influenza or, more whimsically, as the Spanish Lady. That's not because there was anything Spanish about it but because Spain was the first country where it was reported as headline news. In the countries engaged in the First World War, the pandemic was kept out of the newspapers because neither side wanted their opponents to know how badly they were weakened. In neutral Spain, journalists faced no such censorship.

Before it was the Spanish Lady, the pandemic was given different names in different places. British soldiers called it the Flanders grippe, referencing the Belgian region that had become slang for the entire Western Front. Japan was one of the few places to give it a geographically accurate name, where it was the 'American influenza' because they thought it had arrived on ships from the USA rather than because anyone knew what had happened in Camp Funston.

In most locales, the pandemic was named after whoever was most disliked in that locale. It became the German disease in Italy, the

Russian pest in Germany, the Chinese disease in Russia and in newly independent Poland, it was the Bolshevik disease. In much of Africa, it acquired names that meant 'white man's disease' while in Afrikaans, spoken by the dominant white ethnic group in South Africa, it was the 'black man's sickness'[211].

And then the world appeared to forget about it.

Mention the First World War to almost anyone and they will recognise both a monumental tragedy and a pivotal event in world history. More than a century later, the 'Great War' fills pages, screens and museums while academic and armchair historians alike debate machine gun tactics and munitions production.

The 1918 pandemic was the greatest global disaster of the twentieth century, but it left far less of a mark in cultural memory. It's sometimes suggested that it was overshadowed by the First World War but as explanations go, that's less than satisfactory. For one thing, the highest casualty estimate for the four years of the First World War is around 10 million[212], which is only a fifth of the *lowest* estimate of the 50–100 million deaths in the single year of the pandemic[213]. Moreover, not every country was involved in the First World War but as that mass grave in Brevig Mission showed, the pandemic reached into some of the most isolated communities on earth.

In the USA, involvement in the final 18 months of the First World War is vividly remembered and commemorated every Memorial Day, but very few people are aware that more Americans died of the 1918 influenza than in both world wars, the Korean War and the Vietnam War combined[214].

One explanation for why the 1918 pandemic has been so thoroughly forgotten by the historical record may lie in the oft-repeated words of one of the mid-twentieth century's leading historians, Arnold Toynbee, who caricatured his discipline as a description of 'one damned thing after another'[215]. Toynbee was challenging his peers' tendency to see history as a series of events, each of which was influenced by earlier events and influenced subsequent events. Where historians studied people, they focused on the individuals who shaped those events: the political leaders who shaped the destinies of nations and the military leaders who commanded the battles to which conflicting national destinies inevitably led.

An influenza pandemic does not fit easily into such a view. Its origins lay not in great events of human history but, as we'll see, in the history of the influenza virus. Nor were historians able to discern how it affected subsequent events. That may appear strange given that it occurred at a pivotal moment in global history, but it was difficult to untangle its influence from the many other events happening immediately before, during and after it.

To paraphrase Toynbee, there was no damned thing for the pandemic to be after and too many damned things going on at once to see what came after it.

More than 50 years after the pandemic subsided, an English historian called Richard Collier challenged the view that it had ever been forgotten. Collier was an early practitioner of the sub-discipline of social history, which rose to prominence in the 1960s. Social history is less interested in the few presidents and generals to whom statues are erected than in the millions of farmhands, machinists, homemakers and, in 1918, conscripted soldiers whom Toynbee-era historians tended to treat as an anonymous and amorphous mass.

Collier advertised around the world for people to share their memories of the Spanish influenza[216] and found that the memory of the 1918 influenza pandemic had been hiding in plain sight.

He was inundated by more than 1,700 letters sharing half-century-old reminiscences. He shaped them into a social history of the pandemic titled *The Plague of the Spanish Lady*, telling stories such as that of Tersilla Vicenzotto, who roamed the streets of Padua during an artillery barrage to fulfil her husband's dying wish to taste an orange. Of Budapest maidservant Ilona Molnar, who turned blue and expired on a public bench surrounded by a crowd who didn't dare help her. Of the Stockholm reporter Gison Thonfält, who watched two infirmary orderlies bring a letter to a young man's bedside only to find him dead, yet decided to read his mother's admonitions to 'take good care of yourself ... and wrap up when you go out' to his cooling body[217].

The imprint of the pandemic was deepest not in affairs of state or the legacies of great men, which had been the focus of mid-twentieth-century historians, but in the tens of millions of ordinary lives cut short and mourned by hundreds of millions of parents, siblings, children and friends who still remembered them half a century later.

Yet, as Johan Hultin's Alaskan quest showed, the 1918 influenza pandemic made a much deeper mark on medical research than historical research. As the discipline of virology was emerging in the early and mid-twentieth century, the influenza virus was at the centre of it.

What is influenza?

In 1918, doctors could recognise influenza, but they only had a vague idea of what it was. The word was first used during an outbreak in fourteenth-century Florence that was attributed to celestial influence[218]. By the late nineteenth century, medical scientists were looking for a microbiological rather than an astrological explanation and in 1893, Richard Pfeiffer of Berlin's Institut für Infektionskrankheiten* believed he'd found it.

Pfeiffer autopsied several victims of the 'Russian influenza' pandemic that started in 1889, and he found the same bacteria lurking in all of their lungs[219]. Pfeiffer's 'discovery' of the microbe causing influenza was so widely celebrated that the bacterium was known as 'Pfeiffer's bacillus' for several decades, although we now call it *Haemophilus influenzae*.

Pfeiffer got it wrong. *Haemophilus influenzae* is one of those bacteria that infects lungs already damaged by influenza virus. It probably did polish off the people who ended up on Pfeiffer's dissecting table, but it was not the microbe behind influenza.

There was another reason why the 1889 pandemic was a pivotal moment in influenza science: in its immediate aftermath, seasonal influenza outbreaks were recognised as a regular feature of the winter in several European cities[220].

The sudden appearance of seasonal influenza presents a conundrum: did seasonal influenza really spring into existence in the 1890s? Or did the 1889 pandemic focus attention on something that had always been there amid the many causes of coughing that plagued nineteenth-century Europe?

It's one of the many things we still don't know about influenza.

By the beginning of the twentieth century, influenza remained a difficult disease to nail down. In 1912, a Canadian doctor called William

---

* The Institute of Infectious Diseases.

Osler wrote the definitive textbook of its day in which he included a description of influenza that fell short of being definitive. Osler distinguished influenza from other colds not by its symptoms but by its 'extraordinary rapidity of extension and the large number of people attacked'[221]. He was defining influenza by the 'explosive' outbreaks that bring today's healthcare services to their knees. He saw nothing in the nature of the disease to distinguish it from the 'typical grippe', as the common cold was then called.

In the early weeks of the 1918 pandemic, pathologists autopsied the first victims expecting to find Pfeiffer's bacillus in their lungs. Sometimes they found it. More often, they didn't. They often concluded that the victims had indeed died of bacterial infections, but no single bacterial species accounted for the majority of cases.

Medical scientists realised they knew so little about influenza that they didn't even know what they thought they knew.

In 1919, a paper presented to the Royal Society of Medicine stated, 'there can be no doubt that all these bacteria, often acting in conspiracy, have contributed greatly to the recent mortality from influenza'[222], and went on the speculate that scientists had yet to find the primary cause of the disease. The authors were closing in on the truth: most of the Spanish influenza deaths were indeed caused by bacteria[223], but those fatal bacterial infections were a *consequence* of influenza – whatever it might be – rather than *being* influenza.

Meanwhile, two French scientists who shared a given name, Charles Nicolle and Charles Lebailly of the Institut Pasteur's laboratories in Tunisia, were tantalisingly close to the truth. They collected the gunk being coughed and sneezed up by influenza sufferers and passed it through a filter fine enough to remove all bacteria. The filtered gunk caused influenza in both monkeys and in French soldiers who had volunteered as test subjects[224]. The experiment was repeated by researchers in Japan who drew their volunteers from among their laboratory technicians[225].

In both cases, the word 'volunteer' sounds like something of a euphemism and neither group recorded how ill they made their volunteers or even if they all survived. Nevertheless, their results were instructive: both groups showed influenza caused by something small enough to pass through a filter that removed bacteria which, at the time, was the working definition of a virus.

At the time, a virus was such a mysterious and ill-defined entity that many medical scientists saw little difference between a virological explanation for a disease and the sort of astrological explanation that made sense in Renaissance Florence. Many continued to insist that influenza was caused by a bacterium that had not been identified.

The uncertainty would drag on for more than two decades until the early weeks of 1933, when two men had an unexpectedly fortuitous conversation in the National Institute of Medical Research (NIMR) in the leafy London suburb of Hampstead.

Christopher Andrewes's sore throat

One of those researchers was Wilson Smith, whose interest in medical science was sparked during the most miserable years of his life, spent as a Royal Army Medical Corps orderly on the Western Front. After the war, he pursued his interest through a medical degree and went on to work on the embryonic science of virology at the NIMR[226].

His interlocutor was Christopher Andrewes, whose wartime experiences had been less arduous. Andrewes spent most of the war years as a medical student although, such was the shortage of qualified doctors, he was posted aboard a Royal Navy minesweeper before he graduated. He read much of William Osler's textbook, probably including the section describing influenza, sitting on a depth charge. The nearest he came to combat was his ship's only minesweeping operation, when rough seas drove him to his bunk. Andrewes didn't know he'd been in the middle of a minefield until he was safely out of it[227]. After he graduated, he developed an interest in virology and by 1933, it had become the defining interest of his life.

Smith and Andrewes were talking because that winter a particularly bad seasonal influenza epidemic hit London. The Highgate gentility and the Cheapside slums alike were racked with coughing and groaning. Perhaps, Smith and Andrewes thought, all this influenza offered an opportunity to find out what influenza actually was.

Andrewes found himself struggling to concentrate on the ideas they were kicking around. His pulse was racing and the chill of fever was crawling up his spine. Right before Smith's eyes, Andrewes was coming down with the influenza they were discussing. Being a good friend and colleague, Smith urged Andrewes to go home and take to his bed – but

Smith was also a good researcher. He persuaded Andrewes to gargle some saline solution before he left[228].

Like Nicolle and Lebailly 15 years earlier, Smith filtered any bacteria out of the saline. Then he took the filtrate to the NIMR's animal house and sprayed it up the noses of a few ferrets. After a couple of days, the ferrets were starting to look peaky. Their temperatures rose and they developed blocked noses.

The ferrets had influenza.

Smith washed out the ferrets' bunged-up noses, filtered the washings and inoculated it up the noses of some healthy ferrets. The new ferrets developed influenza. He placed flu-ridden ferrets in the same cage as healthy ferrets and, once again, the healthy ferrets got the sniffles[229]. By the time Andrewes was back on his feet, it was clear that the infective agent of influenza was a virus and the first isolate of that virus had come from his throat.

At the time, the only way to sustain a virus in a laboratory was to keep passing it from one animal to another and, because an influenza infection in a ferret lasts for less than a week[230], Smith and Andrewes needed more ferrets than the NIMR's Hampstead laboratory could house.

Smith and Andrewes moved their research to the NIMR's 'farm laboratories' in Mill Hill Village, which was then just north of London and is now just managing to stay aloof from London's suburban sprawl. They continued their influenza experiments in Mill Hill until disaster struck. An outbreak of distemper swept through their ferret cages, breaking the chain of infections that sustained their influenza virus[231].

The fortunes of the NIMR team were nothing if not mixed. One of the last victims of the waning 1933 influenza season was Smith himself who, in turn, gargled saline for Andrewes before going home to recuperate. Andrewes managed to get the 'WS' isolate, as it became known, circulating in yet more ferrets. By the time Smith returned to work, they were back in business.

The virologists of the 1930s were studying a subject they could not see. A virus is much too tiny to be seen through even the most powerful light microscopes, so Smith and Andrewes could only infer the WS virus's existence by watching their ferrets get ill. Nevertheless, sustaining an influenza virus was a major breakthrough and, equally importantly, Smith and Andrewes showed how any researcher could isolate their own influenza.

One such researcher was Frank Macfarlane Burnet at Melbourne's Walter and Eliza Hall Institute of Medical Research. He isolated an influenza virus easily enough[232] but he was less impressed with the ferret-based approach to keeping it going in the laboratory. It was expensive, it was time-consuming and, as Smith and Andrewes had found out the hard way, an outbreak of distemper could send them back to square one.

In principle, Burnet only needed the respiratory tract cells in which the influenza virus replicated. The catch was that the only way to keep those cells alive was to leave them in the ferret.

A modern virologist can order many different types of human and animal cells out of a catalogue along with everything needed to keep them alive in laboratory culture. In the 1930s, there was only one source of living animal cells that could be reliably kept alive in something like culture: a fertilised bird's egg.

An egg is effectively a cell culture system for the cells of a bird embryo. As long as Burnet kept chicken eggs warm in an incubator, he had a much more manageable source of living cells than a cage of ferrets. At the time, virologists were finding that a surprising number of human viruses will infect and replicate in chicken embryo cells and Burnet found that the influenza virus was among them[233].

Burnet's egg-based culture system gave influenza researchers an easy way to work with influenza. During the Second World War, University of Michigan researchers used it to develop a vaccine that was given to some US army recruits[234]. It was a far cry from today's influenza vaccines, partly because they could only produce enough to cover one or two training camps and partly because it simply wasn't very effective[235], but it marked a shift from trying to nail down what influenza was to trying to do something about it.

Global problems, global solutions

The closing months of 1945 were a unique moment in world history. For the first and so far last time, every major power was united in a single alliance. Their accord was founded on the pragmatism of shared enemies rather than mutual warmth and with those enemies crushed, the alliance was already beginning to crumble. Within a couple of years, those powers would be facing off in the Cold War but in those few months of amity, the victorious allies set up the United Nations and the various international frameworks that it encompasses. Eight decades later, the name 'United Nations' often seems more ironic than descriptive but many of those frameworks still enjoy international support today.

One of those frameworks was the brainchild of Christopher Andrewes. Since his sore throat yielded the first influenza isolate back in 1933, Andrewes had been rising through the Medical Research Council hierarchy. By 1945, he was senior enough that while he was establishing the Common Cold Unit, he was pushing for a global system to monitor the influenza virus.

In 1947, Andrewes opened what he called a 'small and unpretentious laboratory'[236] in the NIMR building in Hampstead. He called it the World Influenza Centre, hinting at an objective that was far from small and unpretentious. Under the auspices of what would become the United Nations World Health Organization the following year, Andrewes planned to monitor seasonal influenza around the world and to provide early warning of the next pandemic.

The pretensions of the title proved justified because Andrewes's monitoring system didn't stay small for long. By 1952, it was compiling surveillance from 25 countries[237] from a slightly larger laboratory in a new building on the site of the Mill Hill farm laboratory, where the NIMR had been relocated.

Since then, influenza monitoring has been taken over by the WHO. Under the even more pretentious name of Global Influenza Surveillance and Response System, abbreviated to GISRS and pronounced 'Gissriss'[238], it now collects and compares samples from 129 different countries[239]. The World Influenza Centre is a key part of GISRS and is now in the Francis Crick Institute opposite St Pancras Station.

In 1957, GISRS proved its worth. Its Singapore laboratory identified a new type of influenza that was spreading rapidly and causing much more serious illness than a seasonal strain. It was the first inkling of a pandemic that was about to sweep around the world.

It happened again in 1968, when the Hong Kong centre picked up another new influenza and warned the world that another pandemic was coming.

Both the 1957 'Asian flu' and the 1968 'Hong Kong flu' pandemics killed between one and four million people. That's a lot of deaths, although neither came near the 50–100 million death toll of the 1918 pandemic[240]. The lower death toll was partly because neither the 1957 nor the 1968 virus was as virulent as the 1918 strain[241].

It also helped that by 1957, doctors around the world could treat secondary bacterial infections with antibiotics. They still had no way to suppress the influenza virus itself but if it damaged someone's lungs and let a normally benign bacterium cause pneumonia, those antibiotics were a lot more effective than the hand-wringing and prayer that was all most 1918 doctors could offer.

However, the questions that perplexed influenza scientists in 1918 still perplexed their intellectual descendants 50 years and two pandemics later: where did pandemic influenzas come from and, after their year of global rampage, where did they go?

The answers were hidden in the influenza virus's genome, which scientists only began to explore in the 1970s.

The best place to find an influenza virus, it turns out, is not in some ailing human's throat but in the bowels of a bird. Such 'avian influenza' viruses rarely infect humans directly. Like all viruses, influenza must attach itself to the surface of a living cell before it can insert its genome and start replicating. Each virus has a particular attachment molecule that is specific to a given cell surface molecule. That restricts each virus to the few cell types in the few species that express that virus's target molecule.

For influenza, the exact molecule is called sialic acid, but it must be a particular type of sialic acid. The type in a bird's bowel is different to the type in the human nose and throat but very similar to the type in the human lung. That's good news – up to a point. Good news because the mucociliary escalator* is very good at stopping viruses from getting down the respiratory tract to the lungs. Up to a point, because avian influenza viruses do sometimes make it down to the lungs, which is a very bad place for an influenza infection. The infection triggers inflammation in the same way as a seasonal influenza, but when that inflammation happens in the lungs rather than the nose and throat, it

---

* Chapter 5 contains a description of the mucociliary escalator.

causes a pneumonia that interferes with breathing which is a lot more serious than a cold.

The pneumonia makes avian influenza very dangerous[242] but it's not very infectious. The respiratory tract mucus traps viruses going up the respiratory tract as well as down, so people infected with avian influenza very rarely infect anyone else.

Nevertheless, seasonal influenza viruses do have avian ancestors thanks to an intermediate step that happens in a pig[243]. A pig's upper respiratory tract can host both human and avian influenzas and if two different influenzas go into the same cell at the same time, the viruses that come out are hybrids of the two.

Most of those hybrids are hopeless mash-ups doomed to extinction. It takes a particular combination of functions to stay ahead of the immune response as a virus bounces from cell to cell and human to human. Once in a while, however, a new virus comes out of a cell with a much more dangerous combination of those functions than either of the viruses that went in. That's what happened in 1918, in 1957, in 1968 and probably in every influenza pandemic that we can infer from pre-twentieth-century historical records.

Those pandemic strains are much nastier than seasonal strains, but they don't retain that nastiness for long. As they evolve from recently spawned hybrids into genuine human influenza, they become a little less infectious[244] and a lot less dangerous. That's the answer to the question of where pandemic influenzas go: they're still with us. They are the seasonal influenzas that we all know and despise.

When a pandemic influenza becomes a seasonal influenza, it doesn't stop evolving. Its genome is constantly accumulating mutations, enabling it to infect each of us over and over again because it's a slightly different virus every time we encounter it. Every infection leaves us with immune memory to the influenza that infected us but if we encounter that virus's descendant a year or two later, our immune memory only partially recognises it and it can infect us all over again.

If seasonal influenza viruses are directly descended from pandemic influenzas, it raises the question of why there are only ever three or four seasonal influenza strains circulating at a time. If there are two or three pandemics every century, where are all their seasonal descendants? Thanks to GISRS, we have an answer: whenever a pandemic strain mutates into a seasonal strain, one of the existing seasonal strains fades into extinction[245]. Why? We don't know.

Another unanswered question is why pandemic influenzas lose their virulence. The 1918 pandemic's fatality rate of one in 40[246] is a lot of people from a human perspective, but the virus wasn't about to run out of hosts. Influenza's tendency to lose virulence is often quoted to support the myth that all viruses tend to lose virulence over time – but make no mistake, it is a myth. There's no evidence that loss of virulence is a general trend among viruses and there's even evidence of one of the seasonal influenza strains becoming more virulent in recent years[247].

Amid all the unanswered questions, the constant mutation of seasonal influenza did at least explain why we get so many reinfections with the same strain – or so it was once thought. In recent years, it's emerged that even that is only a partial explanation.

A team led by Jeffery Taubenberger at the American National Institute of Allergy and Infectious Diseases (NIAID) set out to establish how effective immune memory really is against influenza. They recruited seven healthy people who were willing to put up with a cold or two and infected them with a seasonal influenza virus. All seven developed the sort of cold that influenza tends to cause in young, healthy people.

So far, so expected.

Taubenberger left the now recovered volunteers to get on with their lives for a year, then invited them back to be reinfected with the identical virus. That could never happen naturally because of influenza's constant evolution. The only way someone could be reinfected with an identical influenza after a year is if the virus spent that year in a laboratory freezer.

Despite that first infection having given volunteers immune memory to the frozen virus, five of the seven developed another cold[248]. Moreover, Taubenberger's team found the influenza virus happily replicating away even in the volunteers who had not developed colds. Even though they had no symptoms, they were breathing out influenza virus that could infect someone else.

Taubenberger revealed yet another reason why it's so difficult to prevent influenza, and indeed colds in general: the immune system often responds slowly to viruses in the upper respiratory tract even when it has encountered them before. There is a certain logic to that; we breathe in at least six litres of air every minute[249] and that air is laced with dust, pollen, dead cells, other people's respiratory droplets and all sorts of other stuff. If the immune system were too quick to respond to everything that ends up in the upper respiratory tract, we'd feel like we have a cold all the time. Any chronic hay fever sufferer can attest to what that feels like. The price of not having a constant immune overreaction is that the immune system is sometimes slow to respond to viruses in the upper respiratory tract.

The better news is that although immune memory doesn't always keep viruses out of the upper respiratory tract, it's usually pretty good at keeping them out of the lower respiratory tract. That means that an immune response that fails to prevent a cold will usually prevent pneumonia.

The mini-pandemics

After the 1968 pandemic, GISRS reverted to routine monitoring seasonal influenza until May 1977, when public health authorities detected a new strain in the northern Chinese regions of Tientsin, Liaoning and Jilin[250]. It spread as fast as the 1957 and 1968 pandemics, but it was unlike any other emergent pandemic strain because it looked bizarrely familiar.

Its genome looked like a seasonal influenza strain that vanished when the 1957 pandemic strain emerged[251]. In fact, it looked *exactly* like that vanished seasonal influenza, which presented yet another influenza conundrum. How could a strain vanish and reappear after 20 years without having changed in the slightest? If it had been circulating without the GISRS noticing it, it would have evolved into something markedly different to the strain of 20 years earlier.

If it hadn't evolved, it was probably because it hadn't been circulating, which meant it had probably been in a laboratory freezer. That raised an uncomfortable question: how had it got out of that freezer and back into respiratory tracts? The most likely answer was that someone had been either careless or deeply unwise.

At the time, the most likely explanation appeared to be carelessness. If a scientist working with that strain had accidentally infected themselves, they might have infected their friends, colleagues and family without anyone associating the cold going round with that scientist's butterfingers.

In 2004, it emerged that someone might have been deeply unwise. An American virologist called Peter Palese wrote about a study in which the Chinese People's Liberation Army tested an experimental vaccine by deliberately infecting several thousand soldiers[252]. Palese heard about it at a conference in which he got talking to Chi-Ming Chu, one of the Chinese influenza experts who had sounded the alarm in 1977[253]. Palese kept the conversation to himself until after Chu's death because he hadn't wanted to get Chu into trouble for talking about a military project that had never been made public[254].

Most experiments that involve deliberately inoculating volunteers with viruses, such as Jeffery Taubenberger's study on the limitations of immune memory, use a strain that is already in circulation. Deliberately inoculating thousands of people with an extinct strain would have been, as one influenza expert put it to me, 'not so smart'[255]. That's something of an understatement. It would have been impossible to contain an influenza virus in such a large-scale experiment. Today, no research oversight committee worth its salt would approve an experiment that involved infecting one person, let alone thousands, with an extinct strain. Palese's recounting of Chu's story implied that the People's Liberation Army of 1977 laboured under no such constraints.

However, conference bar gossip is at best hearsay. It's not definitive evidence of the origin of the 'red flu', as some American authors described the resurgent strain. A WHO investigation found no evidence of anyone in China working with that strain[256] although, if the military suspected they had accidentally unleashed a previously extinct influenza strain on the world, they might have been less than forthcoming with the WHO investigators.

Wherever that influenza strain came from, it caused an illness that looked far more like seasonal than pandemic influenza[257], which is further evidence that it was a 20-year-old seasonal strain rather than a new pandemic strain. It continued to circulate as a seasonal influenza for the next three decades, when GISRS detected another new strain.

In March 2009, a 10-year-old boy was treated in an urgent care clinic near the Mexican border in California. The clinicians suspected influenza and took samples from his nose and throat. Those samples did indeed contain an influenza virus, but it wasn't one that anyone had seen before. It was a completely new strain and it was already spreading fast, at first on both sides of the Mexico-USA border and then across the world[258]. The first influenza pandemic of the twenty-first century had begun.

Despite the first known case being in the USA, it was dubbed the 'Mexican influenza' or sometimes the 'swine flu' by commentators who did not realise that most, and possibly all, pandemic influenzas originate in pigs. For public health professionals, a new pandemic influenza strain was a nightmare scenario, but it was soon apparent that there was not going to be a repeat of 1918, or even of the less catastrophic pandemics of 1957 or 1968. The disease caused by the newcomer wasn't looking any worse than a typical seasonal influenza[259].

One reason was that this new strain was very similar to the 1918 strain and more pertinently, to the seasonal influenza that had descended from it, vanished, reappeared in 1977 and had been in circulation ever since. Consequently, most people already had immune memory to an influenza virus similar enough to the newcomer to confer some protection against it.

Without the reappearance of the 1918 virus in 1977, nobody born before its disappearance in 1957 would have had any immune memory against the 2009 newcomer. It's impossible to know how much worse that would have made the 2009 pandemic but it's a safe bet that it would have been worse than it was. The world may well have dodged a bullet in 2009 and if the story of the People's Liberation Army vaccination study is true, we may owe a vote of thanks to one of its more unwise projects.

As tends to happen when a new influenza strain appears, the 2009 pandemic strain became a seasonal strain and one of the existing seasonal strains faded into extinction. This time, the short straw was drawn by the descendant of the strain that so mysteriously reappeared in 1977. The direct descendant of the 1918 pandemic strain was finally gone and so far, it has stayed gone[260].

The hunt for the mother virus

In the early 1990s, Jeffery Taubenberger was thinking about how to prepare for the next influenza pandemic. He reasoned that the best way to prepare for the worst was to study the worst, which meant the 1918 pandemic strain at the time of the devastating autumn wave.

By then, he'd moved from the NIAID to the Armed Forces Institute of Pathology in Washington DC. It was the ideal place to ask such a question because the US army still kept many of the autopsy samples its surgeons had taken from soldiers who died of the 1918 influenza.

Seven decades is a long time for a sample to spend in a drawer, but the pathologists of 1918 had intended those samples to last. In the early years of the twentieth century, pathologists adopted the technique of fixing tissue samples in formalin and then embedding them in paraffin wax. It remains the standard way to prepare a sample to study under a microscope because it stabilises every molecule in that sample – at least, so Taubenberger hoped. The technique was never intended to preserve genetic material and nobody had tried extracting viral genomes from 70-year-old samples, but Taubenberger had a laboratory full of the most cutting-edge technology a viral geneticist could want. It was worth a try.

Taubenberger's hunch was proved correct. He found influenza genomes in samples from seven autopsied soldiers[261]. It was a major step forward, but it only took him so far because formalin broke genetic material into fragments before it stabilised it. The army's samples were not going to yield a complete genome.

Then Johan Hultin heard about what Taubenberger was doing.

Since his 1951 expedition to Brevig Mission, Hultin had become more settled, if not exactly sedentary. He'd spent most of his career as a pathologist in San Francisco and amused himself by building an exact replica of a fourteenth-century Norwegian villa and recreating the Minoan labyrinth, site of the mythological battle between Theseus and the minotaur, in the villa's garden. He didn't lose his taste for expeditions; at the age of 60, he was part of a team that made the first ascent of an unnamed 5,844m peak in the Pakistani Karakoram[262]. The first ascent conferred the right to confer a name and the team wanted to call it 'Hultin's Peak' to honour their oldest member. Hultin demurred, insisting on 'Old Codger's Peak' instead[263].

However Hultin amused himself, he never lost his interest in the 1918 influenza. He was on holiday in Costa Rica when he read Taubenberger's paper[264] and immediately understood that Taubenberger needed a complete genome.

He wrote to Taubenberger, offering to revisit that communal grave in Brevig Mission. Taubenberger, who was already thinking about seeking permafrost-preserved bodies, asked Hultin to join an expedition he was putting together.

The 72-year-old Hultin replied, 'I'm free next week.'

That wasn't the timescale Taubenberger had in mind. He was still trying to get the funding together and didn't expect the expedition to leave for another year. To Hultin, that sounded too much like red tape. He trousered his wife's pruning shears and boarded a flight to Alaska.

The Teller village council of 1997 were as willing to help as their predecessors of 1951 and, once again, Hultin set out to exhume the denizens of Brevig Mission. The wooden cross marking their grave had collapsed and their names were long forgotten. Hultin had no idea whose breastbones he was cutting through with those pruning shears. He had particularly high hopes of an obese woman he called Lucy, for want of any other name. There was a possibility that an unusually warm summer might have thawed the permafrost around her but fat tissue, which thaws more slowly than water, might have kept any viral genomes in her lungs stably frozen.

Hultin made sure Lucy and her compatriots were respectfully reburied and, before he left, he exercised his construction skills to build a new cross for the grave.

It was a far quicker flight south than it had been 50 years earlier, and he got his samples to Taubenberger without needing to raid any fire extinguishers. In Washington, a viral genome proved to be a much more attainable subject for a quest than a live virus had been. It took Taubenberger and his team nearly nine years, but they did indeed succeed in obtaining a complete sample of the 1918 influenza virus from Lucy's lung samples[265].

Thanks to Lucy's virus, we know that while the seasonal strain descended directly from the 1918 pandemic strain is now extinct, it has two indirect ancestors that are not. In both 1968 and 2009, it was that now-extinct descendant that hybridised with avian and swine influenzas to give us new pandemics and their seasonal descendants that remain with us today. We're still not completely free of the 1918 influenza, which Taubenberger called 'the mother of all pandemics'[266] – although he called it that before the completely unrelated SARS-CoV-2 emerged.

Hultin took a more philosophical view. In a 2002 interview, he reflected that his 1951 mission never had a chance of success. He couldn't know that even the deep-freeze of permafrost was vanishingly unlikely to leave a virus intact enough to infect a cell. Only when virologists could study a virus's genome directly could they uncover the secrets the residents of Brevig Mission had kept for them.

As Hultin put it, 'The virus sat and waited for me'[267].

Influenza today

Influenza strains may come and go but the disease of seasonal influenza isn't going anywhere. Take a sample of 100,000 people in Britain and in an average winter, it will make between 15,000 and 25,000 of them ill[268]. Most will suffer nothing worse than a few days of the sniffles. but it will send between 45 and 50 of those 100,000 to hospital and kill around a dozen[269]. Most of the deaths will be among people whose age or poor health makes them particularly vulnerable, but that doesn't mean that any of them are ready to check out. Their fatal influenza infection robbed many of them of years of fulfilling life.

However, those average numbers hide the fact that some years are much worse than others. For example, the winter of 2012–2013 saw around 17,000 people die of influenza in England, which has a population of around 55 million[270]. The following winter, 2013–2014, saw a much milder flu season with fewer than 1,500 deaths, but influenza came back with a vengeance in 2014–2015 and killed over 35,000 people. Since then, influenza has killed between 3,000 and 25,000 people in England every year, albeit with a brief respite between 2020 and 2022 when the COVID-19 'lockdowns' also gave us a break from winter flu[271].

With any serious disease, the body count is only part of the story. Viruses virulent enough to kill some of their victims are virulent enough to cause lasting harm to others. One of seasonal influenza's more spectacular manifestations is Guillain-Barré syndrome, which manifests as a paralysis that gets rapidly worse over one or two weeks[272]. With proper hospital treatment, more than nine out of ten sufferers survive and within six months, eight out of ten survivors regain the ability to walk unaided. However, very few fully recover.

Guillain-Barré syndrome is one of many influenza-triggered conditions affecting the heart and the nervous system, including the dysautonomia that afflicts Lucy[273] whose travails were described in the introduction. Quite how a virus that infects the respiratory tract manages to damage the nervous system remains something of a mystery. It's possible that the virus escapes the respiratory tract and infects cells in the heart or the spine. Alternatively, it may cause autoimmune conditions, in which the infection triggers an immune response misdirected against some part of the victim's body rather than against the virus.

How often it causes those conditions is something else to add to the long list of things we don't know about seasonal influenza. It's often less than clear whether influenza was involved in triggering the condition that lands someone in a hospital. Moreover, establishing the percentage of serious cases would require establishing the total number of people infected with influenza, and the systems that monitor how many people get ill with influenza cannot tell how many people are infected without having symptoms[274].

One piece of good news is that we now have ways to prevent and treat serious influenza. Every year, vaccine manufacturers use GISRS samples collected during the southern hemisphere winter to design vaccines used in the northern hemisphere winter six months later. When that winter arrives, samples are taken to design vaccines for the southern hemisphere's next winter.

In most countries, vaccines are given to two groups: children and the vulnerable. Vaccinating children every year builds up a broad repertoire of immune memory against different influenza viruses that doesn't depend on waiting to be infected. Vaccinating the vulnerable, usually defined as the elderly or people whose immune systems can't be trusted to fend for themselves, keeps their immune memory ready to fight the next infection.

A vaccine works by presenting the immune system with all or part of an infectious microbe in a form that can induce immune memory without causing disease. With a microbe as adept at ducking and diving its way around the immune system as influenza, protection that depends on the immune system is always going to be limited. Vaccine-induced immunity is no more effective than the immunity induced by an infection at stopping influenza viruses from causing colds in the upper respiratory tract.

Vaccines are, however, much more effective at protecting against the nastier conditions that happen when the infection spreads beyond the upper respiratory tract[275]. For every ten unvaccinated people who go to their GP and are diagnosed with influenza, only five or six vaccinated people will be ill enough to need a doctor[276]. Those numbers have been fairly consistent since at least the winter of 2015–2016, albeit with one exception. In the northern hemisphere winter of 2017–2018, vaccination gave no protection at all[277]. One of the circulating influenza viruses had evolved so much faster than usual that when it arrived in the northern hemisphere, the immune memory induced by the vaccine – or indeed by previous infections – gave little protection against it. It was a painful reminder that even with a system as sophisticated as GISRS, keeping up with influenza is no easy task.

If influenza lands someone in hospital, either because they were one of the unfortunates who weren't protected by the vaccine or because they were never vaccinated in the first place, there's another line of defence: antiviral drugs that suppress influenza replication[278]. They don't help much with the sort of flu that puts someone to bed for a week, simply because by the time that person has been diagnosed and started taking the antivirals[279] their immune system will already be clearing the virus. Throwing antivirals into the mix will only marginally speed their recovery. However, someone whose immune system has completely lost control of an influenza infection that's running rampant in their lungs is going to be very glad of those antivirals[280].

Seasonal influenza is a bane but it's a predictable one. The next pandemic influenza is far less predictable. The unpredictability lies not in *whether* it will emerge; it inevitably will. It lies in *when* it will emerge and how serious it will be. The 50 years between 1918 and 1968 saw three influenza pandemics, one candidate for the greatest disaster in modern history and two that were less serious in the sense that they killed millions rather than tens of millions of people. As we dodged a bullet in 2009, we've now had more than 50 years since the last serious influenza pandemic. That doesn't mean that the next one is overdue. The hybridisations that give rise to pandemic strains do not follow any particular timetable. They simply happen when a pig happens to inhale the wrong combination of viruses and breathes out their more deadly progeny. The next time that happens may be 50 years from now or it may happen tomorrow.

The world is better prepared for a pandemic than it was in 1918. We have antivirals that suppress the influenza virus, antibiotics for treating the bacteria that exploit the damage it causes and hospitals have a whole range of techniques for supporting people whose lungs are failing. However, COVID-19 showed us that while we're *better* prepared, we're still not that *well* prepared. Health services become overwhelmed by the sheer volume of people needing those treatments, authorities are often slow to impose measures to slow the spread of a virus and at present, it's simply not going to be possible for a vaccine to be manufactured and distributed fast enough to get ahead of a newly emergent pandemic influenza.

All of which was presumably in the mind of Johan Hultin when, at the age of 95, he told a journalist that 'there are only two things that can threaten mankind in the short term. One is an influenza virus, and the other is nuclear war.'[281].

Chapter 7

The hunt for the 'cold virus'

Science and medicine exist on boundaries. Between the known and the unknown. Between health and disease. Sometimes between life and death. Those can be difficult boundaries to navigate, but not because there's any question of which is the right side to be on. Any scientist or doctor who doesn't opt for knowledge, health and life is in the wrong profession. The difficulty lies in working out *how* to make that choice. How to assess the question, the virus or the human being facing them and to identify every factor they need to consider to make the right choice.

That question of how often gives rise to a less clear-cut boundary: what are the limits to what a scientist or a doctor *should* do in their quest for knowledge, health and life and how do they decide what they should not?

One morning in January 1951, a woman was poised on all of those boundaries, though she didn't know that when she hurried through a rainstorm to the Johns Hopkins clinic. She was probably more worried about a different type of boundary. A black woman in Baltimore, or indeed anywhere in the segregated state of Maryland, did not enter a building full of white people without a very good reason. It had taken a year of worsening pain to get her across the boundary represented by the building's entrance.

She signed the form consenting to 'any operative procedures … that they may deem necessary in the proper surgical care'[282] before being examined by a gynaecologist. Part of that examination included a biopsy sample from the source of her pain: the tumour that the gynaecologist found in her cervix. The biopsy was indeed necessary for proper surgical care. Doctors need to know what they're dealing with to decide how to deal with it.

However, she didn't consent to some of that sample being sent to the Johns Hopkins research laboratories where it would be used for research that had nothing to do with her treatment.

If asked, the gynaecologist would have said 'operative procedures' covered the taking of the sample and anything that was subsequently done with it. If it was used for research, that was none of his patient's concern.

Today, that gynaecologist would need to ask for what's called informed consent. He'd start by explaining what samples he's going to take, why he was taking them and what they would be used for. Using a sample for research without telling the patient is a breach of ethics so serious that it can get a doctor barred from ever practising medicine again.

In 1951, that wasn't how things were done. Doctors often tried out experimental procedures on their patients without telling them. In that ethical milieu, the gynaecologist would have seen no reason to ask permission to use one little biopsy sample for research.

He was probably more focused on how to treat that tumour which, the sample revealed, was a very aggressive cancer. He did his best for his patient. He attacked that cancer with everything he had but, in 1951, that didn't add up to much. By the end of the year, she had crossed the boundary from life to death.

Her name was Henrietta Lacks, and she never knew that she had revolutionised modern medicine.

The early days of cell culture

Our current understanding of virology is built on three innovations: the ability to culture a wide range of cells, the electron microscope that enables scientists to see virus capsids in between the cells they infect, and the molecular biology techniques that enable scientists to follow the virus's genetic material inside a cell. Henrietta Lacks's cancer was about to provide the basis of the first of those innovations.

In 1951, the post-war alliance had collapsed into a Cold War that was running hot in Korea, French Indochina* and Malaya, the US Congress

---

* Now divided into Vietnam, Cambodia and Laos.

amended the constitution to limit the presidency to two terms, Dennis the Menace made his debut in *The Beano* and cold research – and virology in general – was at an impasse.

Virologists knew that viruses needed living cells to infect and replicate but as soon as they removed animal cells from an animal, the cells died. Influenza research had been able to progress because the influenza virus would infect ferrets and chicken eggs, but that progress in influenza research showed that influenza viruses only caused a minority of colds.

Research into the non-influenzal colds hadn't made much progress since Alphonse Dochez established that they were caused by viruses*. Nobody had got any closer to establishing the nature of those viruses.

Virologists didn't know it at the time but the species jumping that leads to pandemic influenza strains also made influenza relatively easy to work with. When Smith and Andrewes managed to infect ferrets and Burnet managed to infect chicken eggs, it was because influenza infects a wide range of birds and mammals.

The 'cold virus', as researchers called the mysterious entity behind non-influenzal colds, showed no such promiscuity. The list of animals that researchers tried infecting reads like the inventory of a zoo, ranging from hamsters to hedgehogs[283]. The only success had been with Dochez's chimpanzees, and chimpanzees were too expensive and difficult to get hold of to take the research any further.

As one virologist of that period later put it, 'the best-informed virologist could say "a virus can cause a cold in human beings"; and if he was wise (it would have to be a "he", there were no female virologists studying colds then) he would say no more.'[284]

Such was the problem that Andrewes hoped the Common Cold Unit would solve when he established it in 1946. By then, it was a problem he'd been wrestling with for 15 years.

In 1931, Andrewes visited Dochez's New York laboratory and returned to Britain feeling confident that he could follow Dochez's technique for passing cold viruses between chimpanzees. However, Dochez was supported by an open-handed corporate benefactor.

---

* Chapter 1 describes how Dochez showed that colds are viral infections.

Andrewes could find no one willing to pay for a chimpanzee house in Britain.

It was a problem that suggested its own solution. Dochez had resorted to chimpanzees as a substitute for humans so, if Andrewes couldn't afford chimpanzees, perhaps he could keep things simple – and cheap – and recruit humans instead.

He took his idea to Henry Dale, who was then the director of the Medical Research Council, suggesting that students and the unemployed might be willing to endure a cold for a small remuneration. Dale baulked at the idea of recruiting the unemployed, who might be induced to put their better judgement aside by the offer of cash[285].

Dale's reservations placed him ahead of his time, at least in his views of medical ethics. A fundamental principle of today's medical ethics is that a volunteer can only be compensated for their expenses, and they should not be paid so much that they might be incentivised by the cash alone. No such principle guided the laissez-faire medical ethics of the 1930s, when researchers were left to decide what was acceptable for themselves.

Dale did allow Andrewes to recruit medical students training at St Bartholemew's Hospital as long as he didn't pay them. Andrewes quipped that 'the next best thing to a chimpanzee is a Bart's* student'[286] and set about infecting students in the same way that Dochez had infected his chimpanzees. To Andrewes's disappointment, his quip proved more correct than he'd anticipated. His research ran into the same difficulties that Dochez had used chimpanzees to avoid.

Medical students in the middle of London were constantly surrounded by people exhaling cold viruses[287]. If an inoculation didn't cause a cold, Andrewes had no way of knowing whether it was because there was no virus in the inoculation or because the student had been recently infected with that virus and was immune to it. Meanwhile, another student who had been inoculated with a control solution containing no virus at all might come down with a cold that they'd picked up in a lecture theatre, hospital ward, tram, pub or anywhere else someone might have breathed it on them.

---

* Bart's was and remains the affectionate nickname for St Bartholemew's Hospital.

When it came to cold research, medical students were next best to a chimpanzee by a very long way.

Andrewes's Bart's experiments got him no closer to getting a handle on the 'cold virus' and a couple of years later, his and Smith's isolation of the influenza virus refocused his attention. However, Andrewes never lost interest in the common cold and in 1946, he was once again knocking on an MRC director's door. By then, he was senior enough to pitch the much more ambitious plan for an entire research institute dedicated to the 'cold virus'.

Volunteers in the unequal struggle

We can't know what was in Andrewes's mind when he first conceived the Common Cold Unit, but the influences of Dochez's chimpanzee house and his abortive experiments with medical students are plain to see. The CCU was designed to enable its researchers to control and isolate human volunteers in the same way that Dochez had controlled and isolated his chimpanzees.

The CCU began its quest for the elusive 'cold virus' on Wednesday 17 July 1946, when the first group of volunteers arrived at Salisbury station and were driven to Harvard Hospital. They were the first to experience a routine that would change little until the CCU was closed in 1990.

The first order of business was to check that the volunteers were not already harbouring an infection. A clinical examination and a chest X-ray ensured nobody had tuberculosis, which was widespread in 1940s Britain. However, the only way to be sure that nobody had brought a cold with them was to wait three days and see if they developed any symptoms.

With the waiting began the isolation. Volunteers were allocated to huts in pairs. Some arrived with friends while others arrived alone and were allocated a hut-mate for their time at the CCU. Volunteers were asked to stay at least 30 feet away from anyone other than their hut-mate for the duration of their stay, although they could take walks or cycle rides in the surrounding countryside as long as they maintained that 30-foot distance. They were issued with gauze facemasks which they donned if they encountered an oncoming group of walkers on a narrow path. One of the many matrons the CCU employed over its 44 years briefed them that if they met a friendly stranger, they should slap handkerchiefs over their faces and shout, 'Unclean! Unclean!'[288]

Such theatrics were rarely necessary. The residents of Salisbury knew of the CCU's work and usually followed the 30-foot rule without being asked.

The experiments began on Saturday and usually involved dripping saline containing a cold virus up the volunteers' noses, although volunteers could find themselves subject to all sorts of uncomfortable elaborations like the soggy socks experiment described in Chapter 4. Nevertheless, there was never any shortage of volunteers, some of whom enjoyed their time at the CCU enough to come back. One schoolteacher volunteered six times in 11 years.

Some reminiscences paint a picture of holiday hijinks reminiscent of the sitcom *Hi-de-Hi!* One male guitarist, who volunteered nine times, spent much of one of his stints playing duets with an attractive oboist at the requisite 30 feet of separation. On another occasion, two men and two women found themselves in flats at opposite ends of the same hut and broke through their ceiling panels so they could pay sneak visits to each other[289].

When I spoke to James Cherry, who worked there in the 1960s, he suggested that such fun and games were not the only reason why so many were willing to volunteer; in those years of post-war austerity, many people could not afford holidays so the risk of a cold was a small price to pay for a free two weeks in the Wiltshire countryside[290].

When overseas travel was beyond the reach of any but the wealthiest and summer holidays were often taken in prefabricated huts assembled into 'holiday camps', the rather dismal demeanour of the CCU was probably less off-putting than it would be to a modern holidaymaker. Moreover, the surrounding countryside was pleasant, the chefs were able to work wonders with the ingredients they could get by pooling the volunteers' ration cards and in the winter, the efficient heating systems[291] were still something of a novelty.

Andrewes made the most of the rural setting to indulge his hobby of entomology. He often took his junior researchers on long walks in the country, punctuating discussions of virology with pauses to capture insects that caught his eye. James Lovelock later recalled being told to walk in front of Andrewes to flush out sawflies[292], for which Andrewes had a particular passion.

In those early years, the CCU's success in attracting volunteers was not matched by any success in isolating cold viruses. Andrewes and his colleagues had little difficulty in passing a cold virus from one volunteer to another but there was only so much they could do while the only cells the 'cold virus' would infect were in a volunteer's respiratory tract. They really wanted to work with viruses in the laboratories set up next to the volunteers' huts, but the 'cold virus' refused to cooperate.

With the benefit of hindsight drawn from modern virology, it was always going to be incredibly difficult to culture a virus that replicated in living human cells without a reliable technique for culturing living human cells. They did their best with nutrient broths concocted from ingredients from local farmyards. Amniotic fluid collected when cows gave birth was mixed with chick embryo extract and often a dash of human blood. It was messy, it was inconsistent – one cow's amniotic fluid is constituted differently from another's – and even when a broth kept cells alive, it didn't keep them alive for very long. One laboratory manual summed up the vagaries of mid-twentieth-century cell culture by opening a chapter with the 'double double toil and trouble speech' from *Macbeth*[293].

Finding large numbers of human cells was no picnic either. The CCU's cell culturists mostly worked with lung cells harvested from aborted foetuses. Their American colleagues had an even harder time of it because of the USA's much more restrictive laws controlling abortion. An alternative option was usually cells from kidneys that were removed during a complicated procedure to treat hydrocephalus, in which spinal fluid becomes trapped inside the skull and must be drained to avoid compressing the brain.

Culturing cells was difficult. Culturing viruses inside those cells added another layer of complication and, for the CCU scientists, enduring frustration. With no way to tell whether a virus was replicating in a culture by looking at it, all the CCU scientists could do was take a nasal secretion from a cold-ridden volunteer, filter out anything larger than a virus, inoculate a culture, filter it again and try to infect another volunteer with it. Again and again, they tried and failed. Volunteers inoculated from cell cultures enjoyed their country hikes and their games of table tennis with sanitised bats blissfully untroubled by the colds that the scientists were trying to inflict on them.

In April 1951, Andrewes dolefully reported that 'after 6½ years' failure ... we were almost prepared, last January, to give up the unequal contest'[294]. The word 'almost' pointed to a slight ray of hope. They had managed to induce colds in two volunteers from a virus cultured in cells harvested from an aborted foetus[295], but they were still nowhere near what they could do with influenza.

Then Henrietta Lacks walked into that Baltimore clinic.

The birth of HeLa

While Henrietta Lacks was reeling from the news that she had cervical cancer, her cells were arriving in George Gey's laboratory[296]. Gey described himself as a 'vulture'[297] who would get hold of cells from wherever he could get them and try to culture them. He wanted cells that would divide, divide and divide again without dying, but after three decades of vulturing, such immortal cells seemed an impossible dream. The best he could do was to keep a few cell types alive for a few days but inevitably, he would watch the cultures deteriorate until the cells died.

Dividing, dividing and dividing again is exactly what cancer cells do. Most cell types have a built-in ageing process that limits how many times they can divide before they die. If the processes that impose that limit break down, the cell may divide repeatedly to spawn a tumour of identical and similarly uninhibited progeny cells. It was the last thing that Henrietta Lacks wanted in her cervix and exactly what George Gey needed in his laboratory. He watched the cells he called HeLa, an abbreviation of Henrietta's name, turn from that tiny biopsy into an ever-expanding culture of identical cells without showing a single sign of deterioration. They're still dividing in cell cultures across the world.

Today, if a scientist at Johns Hopkins, or any other university, came up with something as groundbreaking as the HeLa cell line was in 1951, the university would slap a patent on it and try to commercialise it. George Gey was working in a different time. In the 1950s, American academics firmly believed that it was unethical to commercialise their discoveries[298]. The ethics of intellectual property were then much clearer than the ethics of patient consent.

It probably never occurred to Gey that the HeLa cells might be valuable intellectual property. He didn't even publish a paper on how he established the HeLa line. His only description was in a presentation to the 1952 American Association for Cancer Research, which leaves us with no written record of exactly what he did beyond a rather vaguely worded abstract of less than 150 words[299].

It was typical of Gey, who preferred to focus on the next thing rather than on writing up the last thing[300], but he was happy to hand out samples of HeLa cells to anyone who asked for them.

Among the researchers who asked for them were those hunting the elusive 'cold virus'.

Patient 67

In the winter of 1952–1953, there was a cold going around Fort Leonard Wood, a US Army training base in Missouri. A cold wouldn't do any lasting harm to the fit young recruits but it disrupted their training, which is why the US army was engaged in the same unequal struggle as Christopher Andrewes and his colleagues across the Atlantic in the CCU.

At Fort Leonard Wood, Maurice Hilleman and Jacqueline Werner of Washington DC's Army Medical Service Graduate School* were doing the struggling. They spent that winter examining one sniffling soldier after another.

In what they called the 'mid-portion' of the outbreak, Hilleman and Werner were able to isolate and identify seasonal influenza. However, there were plenty of cold-ridden soldiers to examine before and after the few weeks of seasonal influenza's annual visitation. They realised that they were dealing with overlapping outbreaks of seasonal influenza and the elusive 'cold virus'. As Andrewes and his colleagues had found at the CCU, the latter caused an indistinguishable illness but refused to replicate in chicken eggs or any of the animals the influenza virus readily infects[301].

A year or two earlier, that would have been as far as Hilleman and Werner could have taken their research, but now they had another trick to try: the immortal HeLa cell line.

---

* Now the Walter Reed Army Institute of Research.

The soldier whose cold broke the two-decade impasse in cold research is recorded only as Patient 67, which suggests that Hilleman and Werner tried and failed 66 times before they got a result worth publishing. They washed out Patient 67's sore throat, filtered out anything larger than a virus and dripped the resulting filtrate into a culture of HeLa cells. The results were striking. Those immortal cells that, given sufficient nutrients, would divide eternally without showing a sign of ageing, turned dark and shrivelled up. When Hilleman and Werner harvested some of the culture medium and inoculated a fresh batch of HeLa cells, they too shrivelled up and died.

Something in Patient 67's throat was killing HeLa cells and it wasn't influenza.

That was encouraging but after so many researchers had failed to isolate the 'cold virus', Hilleman and Werner wanted another layer of confirmation before they were sure they'd succeeded. They reasoned that if those HeLa cells were being killed by the same virus that was causing Patient 67's cold, then Patient 67's immune response would have flooded his bloodstream with antibodies against that virus. Sure enough, when they mixed culture medium from dying cells with Patient 67's serum – the liquid fraction of his blood – the next batch of HeLa cells carried on dividing as if nothing had happened. Patient 67 had immune memory to the virus that had made him ill.

Hilleman and Werner washed out more soldiers' throats and from two of them, they isolated the agent that killed HeLa cells. Only then did they feel confident that they'd isolated the 'cold virus' from the soldiers of Fort Leonard Wood. It opened a whole new chapter in the study of cold viruses which, for most scientists, would be the crowning achievement of their career. Maybe it was for Jacqueline Werner, although it's hard to tell. Her name vanishes from the publication record after a few more papers on the same subject.

For Maurice Hilleman, it was merely a prelude to a career so eventful that many articles describing it don't even mention his pioneering cold research. Hilleman went on to head up the vaccine division at the pharmaceutical firm Merck, Sharp & Dohme where he led the development of so many of the vaccines in use today that one headline writer dubbed him 'the man who saved your life'[302] which, for some readers of the book, he undoubtedly is.

From the cold virus to the cold viruses

The HeLa cell line revolutionised virology by solving the problem that had bedevilled the early years of the CCU. Virologists could finally study viruses in their laboratories and, even if they still couldn't see the viruses directly, they could directly observe how they affected the cells in which they replicate. It enabled them to do far more than they could ever achieve by simply passing viruses between eggs or laboratory animals, let alone human volunteers.

For virologists hunting the 'cold virus', it was particularly fortuitous that HeLa cells were derived from the epithelium, which is the thin layer of tissue that covers the external surfaces of the body inside and out, including the skin, the intestines, the respiratory tract and, among other things, the cervix. Ideally, a cold virus researcher would want cells from the respiratory tract epithelium, which is where cold viruses replicate, but cells from the cervical epithelium were almost as good. They were similar enough that at least some of the cold viruses – including the adenoviruses that Hilleman and Werner isolated in Fort Leonard Wood – will infect and replicate inside them.

Once Hilleman and Werner showed that cold viruses could be cultured in HeLa cells, many other virologists in many other institutions used their technique to isolate viruses from up blocked noses. When they exchanged samples and compared notes, they realised that their successes did not mark the end of the hunt for *the* cold virus but the beginning of the hunt for *many* different cold viruses. By the mid-1950s, enough cold viruses had been isolated by enough research groups that when a new publication described a new isolate, it was often difficult to tell whether they were describing something entirely new or something being researched elsewhere under a different name.

The only solution was guaranteed to make any researcher's heart sink: they had to form committees.

One such committee established that Hilleman's and Werner's virus belonged to the same group as a virus that had been isolated a year before by a research team at Johns Hopkins University who, ironically, had not used the HeLa cell line established in their own institution. The Hopkins team's characterisation was so vague that they weren't even sure they were working with a virus and not a bacterium[303] but the committee based the name of the new group of viruses on a feature of that earlier isolation.

That Hopkins team had isolated their virus from a child's swollen adenoid gland. The adenoid is one of the tonsils, located at the back of the nasal cavity, just above the gap between the palate and the throat. In the 1950s, removing the tonsils was a routine operation that was the usual response to any sort of tonsillar swelling. Today's ear, nose and throat specialists tend to regard their predecessors as being rather too quick to reach for the scalpel but it made for a ready supply of discarded tonsils for any researchers who were interested in them and, when that committee put their heads together, it suggested a name for the new group: the adenoviruses[304].

There is something fitting about the first cold virus to be isolated and cultured belonging to the same group that gives us the oldest known cold virus: the adenovirus that would later be identified in the Yana River teeth preserved from over 30,000 years ago*.

Here I must apologise, on behalf of my profession, for the etymological abominations that my colleagues conjure when they form committees. In writing this book, I've tried to keep the technical terms as light as possible, but I have to ask you to bear with me. There's no getting away from what the viruses are called.

While humanity is plagued by cold viruses, virology is plagued by ugly acronyms and awkward portmanteaus of Greek, Latin or both. Moreover, names that made sense to the committees that coined them often make less sense in the light of later research.

The adenoviruses are a case in point; we now know of more than 80 human adenoviruses[305], many of which go nowhere near the adenoids. Others infect the intestines and cause gastroenteritis or, as most of us call it, stomach bugs, stomach flu or simply the shits. Others infect the membrane covering the eye and cause conjunctivitis. Nevertheless, the family is stuck with the name 'adenoviruses' because the first of its members to be isolated happened to be infecting an adenoid.

Around 50 adenoviruses find their homes somewhere in the upper respiratory tract and may cause a cold until the immune response gets its act together to kick them out – or they may not. Sometimes they

---

* Chapter 2 describes the long journey of those teeth and the viruses they carried.

manage to avoid attracting the immune system's attention and carry on replicating for a much longer period, which may cause the sort of swelling that once led to children having their adenoids removed from their throat and given to a laboratory team – or they may cause no symptoms at all[306].

The adenoviruses only account for between one in ten and one in twenty of all the colds we get – unless you happen to be a military recruit. As Hilleman and Werner built on their original success, they found that three of those adenoviruses had a particular affinity for military training bases. Every winter, those adenoviruses reduced recruits to miserable wrecks in their bunks and, much worse from the perspective of a drill sergeant, disrupted carefully planned training schedules. As unpleasant as the outbreaks were for the recruits in basic training, those recruits were left with immune memory to the adenoviruses that caused them. Those same outbreaks rarely affected their instructors, who retained immunity from their days as raw recruits[307].

Hilleman and Werner were seeing a pattern of disease that often happens when groups of young people are packed together in close proximity: if one of those young people brings in an airborne infection, it spreads like wildfire. What happens next depends on how effective the immune memory it leaves behind is and, in the case of adenovirus, that immune memory tends to be pretty effective. The instructors who had themselves been infected as recruits remained unaffected by the adenoviruses their recruits projected at them with a hoarse 'Sir, yes sir!'

Nose poison

In 1957, David Tyrrell arrived at the CCU and joined the hunt.

The CCU hadn't been his first choice but in the 1950s, first-choice posts went to men who had spent time in uniform. Tyrrell had fancied the Royal Navy, but his bad eyesight made him a reluctant exemptee from national service. Nevertheless, he landed a post at New York's Rockefeller Institute where he learned the latest techniques in cell culture. After three years at the Rockefeller, he brought those techniques back to Britain when he got a job at his alma mater, the University of Sheffield[308].

When his Sheffield contract expired and he was offered a post at the CCU, he didn't feel able to refuse. In 1957, he drove his family to Salisbury where, he later reminisced, 'even the sunshine could not remove our feeling of dismay'[309] at his first sight of Harvard Hospital. He was among the many who thought it looked more like a prison camp than the ocean liner of Andrewes's imagination.

Nevertheless, Tyrrell soon accepted that he'd made the right choice for both himself and the CCU. He had everything he needed to apply the cell culture techniques he'd learned in New York while the CCU was sorely in need of someone with those skills.

After years of the 'unequal struggle' that Andrewes bemoaned, Tyrrell managed to get a virus into culture. He used short-lived embryo cells rather than the immortal HeLa cells, cultured using the CCU's improvised nutrient broth scraped together from farmyard waste. Despite the vagaries of the technique, Tyrrell got it to work. He was able to pass this new virus between cell cultures and volunteers, enabling Andrewes to finally report to the MRC board that the CCU had succeeded in isolating a cold virus.

Meanwhile, Winston Price at Johns Hopkins had used HeLa cells to isolate a cold virus he called the JH virus after his institute[310]. Price and Tyrrell exchanged samples and notes and established that neither the JH virus nor Tyrrell's isolate was inactivated by ether but both were inactivated by a slightly acidic solution. Both viruses killed the cells they replicated in. Putting those together, they concluded that they were working on very similar viruses that were very different to the adenoviruses.

Their next move was to see whether they had independently discovered the same virus. They found that an infected person produced antibodies that neutralised whichever of the two they had been infected with, but not the other one. Tyrrell and Price had isolated different viruses but the similarities were so strong that they concluded they had simultaneously isolated two different members of a hitherto unknown virus family.

It turned out to be an important family. They had discovered the virus group that causes between a third and a half of all colds[311], making it the most prolific group of cold viruses.

Price and Andrewes called them the ECHO viruses, which was an acronym for 'enteric, cytopathic, human, orphan'[312]. Enteric because they were – somewhat confusingly – lumped together with a closely related group of viruses that cause gastroenteritis rather than colds, cytopathic meaning they killed the cells they replicated in, human because they stubbornly refused to infect any laboratory animal other than a human and orphan because cold symptoms were so nebulous that they could not be associated with a specific disease.

In 1963, they were renamed the 'rhinoviruses'[313]. The term combines the Greek word for nose, 'rhino', with the Latin 'virus', literally meaning 'poison', which always was misleading; an infectious agent that replicates itself is categorically different to a chemical poison.

The committee that coined the word 'rhinovirus' decreed that most of our colds are caused by 'nose poison', which certainly feels appropriate. The members did not appear to realise that they'd made it impossible to discuss cold viruses without thinking of a rhinoceros.

Once virologists knew to look for rhinoviruses, they found a lot of them. It quickly became apparent that there was not one rhinovirus but many. The first indications of rhinovirus diversity came from the approach Tyrrell and Price used to establish they had discovered two different viruses: someone infected with a rhinovirus would develop antibodies that protected them against that rhinovirus and neutralised it in culture, but had no effect on any of the other rhinoviruses.

Researchers shared samples of serum containing antibodies known to neutralise a given rhinovirus and when they isolated a new rhinovirus, they tested it against those samples to see which one neutralised it. If none of those samples neutralised it, they'd got a new one.

Microbes classified by the antibodies that neutralise them are called serotypes because antibodies circulate in the liquid fraction of blood called the serum. In the decades following Price's and Tyrrell's simultaneous discovery, serotype after serotype was discovered. By the time the serotyping system was replaced with a system based on the genetic sequence of the rhinoviruses in 2020, there were 165 known serotypes and at the time of writing, there are probably more still to be found[314].

The massive diversity of the rhinoviruses is the reason why they cause so many colds. When we catch one of them, we become immune to that one, but only that one. The other 164 are still out there. As we progress through life, we accumulate immunity to the many different rhinoviruses one sniffle at a time, but we're unlikely to live long enough to become immune to them all.

As colds go, most rhinovirus infections are relatively mild. They rarely spread beyond the cells lining the nose and the throat and unlike influenza, they don't kill those cells[315]. However, every cold virus occasionally causes something more serious than a cold, and although only a very small proportion of rhinovirus infections cause serious illness, the sheer number of rhinovirus infections means that that low proportion adds up to a substantial number of cases. One American study looked into what was hospitalising children under five years old with breathing problems and found that a quarter of them were infected with rhinovirus[316].

Most of those children had already been diagnosed with asthma, making it look like the rhinovirus infection was exacerbating an underlying problem. However, there is some evidence that rhinovirus may be the root cause of at least some cases of asthma. We know that thanks to the researchers and volunteers of the University of Wisconsin's Childhood Origins of Asthma study. The COAST study, to use its abbreviation, followed over 250 children from the day of their birth until their sixth year. The researchers found that if a child had a rhinovirus infection bad enough to make them wheeze by the time they were three years old, they were ten times more likely to have asthma by the time they were six[317].

That's not conclusive proof that rhinovirus causes asthma. The children in the COAST study were recruited because at least one of their parents had either asthma or an allergy to something airborne like mould spores or windblown pollen[318], which meant that the children were always at high risk of developing some sort of respiratory problem. It's quite possible that the condition that would develop into asthma was there before they had their rhinovirus-induced wheezing episode, even if it had yet to be diagnosed.

There is one piece of evidence that points the finger of blame in the general direction of rhinovirus, if not directly at it. If the wheeze was caused by a rhinovirus infection aggravating a problem that was already there, then wheezing caused by any other cold virus would be just as likely to be followed by a diagnosis of asthma.

The COAST researchers explored the possibility by comparing rhinovirus with another cold virus, the respiratory syncytial virus. They chose RSV because, as we'll see in Chapter 8, it's far more likely to cause serious illness in young children than the rhinoviruses. If there was a general association between cold viruses and asthma, the COAST researchers reasoned, it would be at least as likely to follow an RSV infection as to follow a rhinovirus infection.

That didn't happen.

Despite an RSV infection being far more likely to send a child to hospital than a rhinovirus infection, children who suffered an RSV-induced wheeze before they were three were no more likely to develop asthma than children who had never suffered RSV-associated wheezing. The association was specific to wheezing caused by rhinoviruses.

The COAST researchers suggested that those rhinovirus infections may be causing 'airway remodelling'[319]. They were referring to the way that young children are rapidly growing and developing, which means that all of the structures of the human body are growing and developing inside them. Perhaps a rhinovirus infection hitting them at that critical point affects the way their respiratory tract grows and develops, pushing it toward the thicker and overly responsive tract lining that is typical of asthma[320].

At present, the suggestion that rhinovirus-induced wheezing causes asthma is a hypothesis rather than a theory. The mechanism that connects one to the other is unknown, which makes it impossible to be certain that the rhinovirus that made a child wheeze also caused their asthma. It's at least possible that wheezing is more likely when rhinovirus infects a respiratory tract that's taken a developmental wrong turn that will later lead to asthma – although if rhinovirus-induced wheezing is merely an early sign of asthma instead of a cause of it, it raises the question of why RSV-induced wheezing isn't such a sign.

At the time of writing, that's as far as the research has got. Other studies have also found an association with rhinovirus wheezing and asthma[321] but none have gathered anything like the level of detail as the COAST study. Given how difficult it is to persuade funding bodies to sustain a study that follows a decent number of children through childhood, it's unlikely that there's going to be another study that builds on the COAST findings anytime soon. It leaves us in the unsatisfactory position of saying that there's pretty good evidence that rhinovirus infection causes asthma, but we don't know how likely it is to happen or even whether it happens at all.

The legacy of Henrietta Lacks

George Gey and Henrietta Lacks never met but between them, they galvanised cell culture techniques which, in turn, galvanised virology. The HeLa cell line played a pivotal role in ending the two-decade struggle to identify the 'cold virus' by showing that a cold was not a single disease with a single cause but rather a set of symptoms caused by many different viruses.

Moreover, it made it much easier to study and characterise those viruses. A mid-twentieth-century virologist would start with the battery of chemical tests that showed Tyrrell and Price that they had not isolated adenoviruses but something altogether new. Then he would identify the virus's serotype by testing it against a library of different sera to see which contained the antibodies that inactivated it. By modern standards, the process of characterisation was time-consuming and cumbersome, but it was enormously easier to do all that using a laboratory cell culture than a bunch of human volunteers whom he may or may not be able to infect with that virus.

It's simple enough to describe the scientific advances that followed from that biopsy of Henrietta's Lacks's cervical tumour. The ethical question is less straightforward.

When Gey and his colleagues chose knowledge, health and life for the future, did they make the right choices for Henrietta Lacks in the present or, perhaps more pertinently, should they have made those choices for her at all without asking her?

Nobody told Henrietta Lacks that anyone was going to try to culture her cells but at the time, doctors very rarely told their patients whether they were involved in research. The gynaecologist who took that biopsy certainly couldn't have anticipated that her cells would be thriving in laboratories around the world seven decades later; at the time, nobody had kept cells alive in culture for more than a few weeks.

George Gey certainly thought he was behaving ethically when he gave HeLa cells to anyone who asked for them, but he probably didn't see the intellectual property conundrum he was creating: several biotechnology companies have since developed and patented products that use HeLa cells so, while the cells themselves are what we might now call open-source, they are part of products that are sold for a profit. Those profits were being raked in for decades before anyone even thought to tell Henrietta Lacks's children that there was such a thing as the HeLa cell line.

Discussion of those ethics is complicated because informed consent is not the only ethical principle that has changed. If today's ethics were being followed in 1951, nobody would know the name of Henrietta Lacks. Being involved in research, with or without consent, does not waive someone's right to keep their medical information confidential. To name a sample after a patient, as Gey did when he called the cell line HeLa, would be considered a violation of that confidentiality. None of the millions of people who now know Henrietta Lacks's name have any right to know what was going on in her cervix in that last year of her life but, because her confidentiality wasn't respected, it became public knowledge.

In the toolbox of modern cell culture, the immortalised cell lines called WI-38 and MRC-5 are as important as HeLa. Both are used for non-profit research and to produce commercial products and, like HeLa, both came from someone. Unlike the HeLa cells, nobody has any idea who, because their donors' confidentiality has been protected. The names WI-38 and MRC-5 refer to the institutes they were first cultured in, not the individuals they were cultured from.

If Henrietta Lacks's story were purely a tale from the past, it could be a morality tale told to trainee doctors and researchers. Never do a research procedure without informed consent, the tale would conclude, and always protect the confidentiality of the person who gives you that consent.

Yet Henrietta Lacks's story continues in the present. Her cells thrive in laboratories across the world and continue to be used in many different applications. Some of those applications are of great benefit in modern medicine. Some are money-spinners for the corporation that owns the laboratory. Some are both.

As Henrietta's daughter, Deborah Lacks, put it decades later, 'She helpin' lots of people. I think she would like that ... But I always have thought it was strange, if our mother cells done so much for medicine, how come her family can't afford to see no doctors? Don't make no sense. People got rich off my mother without us even knowin' about them takin her cells, now we don't get a dime.'[322]

Sneeze

How come indeed?

It would not be the last time that we might wish that cold researchers had navigated that ethical boundary with a little more aplomb.

Chapter 8

Robert Chanock joins the hunt

When Robert Chanock graduated from medical school in 1947, he believed something that no doctor or scientist should ever believe: that everything his profession needed to know was known. He didn't fool himself that *he* knew it all – Chanock was never hubristic – but he believed the professors of his alma mater, the University of Chicago, had the answer to every question that medical practice might ever need answered[323].

His illusions were shattered when the daughter of his immediate superior was rushed to the hospital with as bad a case of croup as anyone had seen. The girl's throat was so inflamed that she could hardly breathe. Chanock and his colleagues had to pass a tube down her throat so a ventilator could pump air in and out of her lungs. He later said, 'it was touch and go. She could have died at any moment.'[324]

Treating a child with life-threatening croup was not an unusual experience for a junior doctor, but what really shook Chanock was that none of his mentors had much idea what caused it. All he could find out was that the causative agent of croup was generally thought to be a virus but, because nobody had managed to isolate it, nobody knew for sure.

Chanock was finding out that research into the hypothetical 'croup virus' was stuck in much the same place as research into the 'cold virus'.

The croup virus

Chanock became a doctor almost by accident. He'd wanted to study biology, but the Second World War saw him drafted into the army as soon as he finished high school. The army was so impressed with his test scores that they offered to send him to medical school. It wasn't biology, but a free medical degree sounded better than the infantry[325].

He never saw a battlefield operating theatre because the war ended before his medical degree. The army released him into civilian practice where that little girl's croup opened his eyes to the yawning knowledge gaps that the medical profession struggled with. He started thinking about rekindling his interest in biology through medical research.

He joined one of America's leading virologists, Albert Sabin, at the Cincinnati Children's Hospital. Sabin would later become famous as the inventor of the oral polio vaccine, but when Chanock met him in 1950, Sabin was better known as the boss from hell. His junior researchers often hid in cupboards to escape his temper tantrums[326] but Sabin took a shine to Chanock. On one occasion, Chanock recalled that he 'screwed up terribly' and instead of reaming him out, Sabin threw an arm around his shoulders and said, 'Bob, don't be depressed, you know, I made a mistake once.'[327]

Sabin's idea of reassurance was less than helpful, but he did support Chanock's quest for the croup virus, at least until it was interrupted by the Korean War. That free medical degree enabled the army to redraft him and he spent a year in a military research institute in Japan. As soon as he was discharged, he returned to Cincinnati to return to his hunt for the 'croup virus', making him the only researcher who ever left Sabin's department and dared to go back.

Chanock's army service had coincided with the advances in cell culture covered in the last chapter and while other virologists applied those advances in hunting the 'cold virus', Chanock used them to close in on the 'croup virus'. He swabbed the throats of children with croup, filtered the samples and by 1954, he could consistently isolate a virus that would replicate in cultures of monkey kidney cells[328]. He initially called it croup-associated virus, but it soon emerged that only a small number of people infected with the virus develop croup. The vast majority simply got a cold.

The 'croup virus', it turned out, was another type of 'cold virus'.

It was only a matter of time before Chanock's croup-associated virus ended up on a naming committee's agenda where it became the parainfluenza virus or PIV. The Greek prefix 'para' means 'next to', so 'parainfluenza virus' refers to it causing a disease similar to influenza.

Like the rhinoviruses and adenoviruses, there proved to be more than one serotype of the parainfluenza virus, although it is nowhere near as diverse as either of them. There are only four known parainfluenza virus serotypes, which is far fewer than the 80-odd adenoviruses, let alone the more than 160 rhinoviruses. Between them, the four parainfluenza viruses cause around one in 20 of all colds in both adults[329] and children[330]. Adults get fewer colds overall, so that does mean that adults get fewer parainfluenza virus infections, but we're never completely free of them. The immunity we get from an infection tends to be incomplete and short-lived[331], so we can expect the occasional reinfection throughout our lives.

However, Chanock's original name of croup-associated virus was apposite. The parainfluenza viruses are the most likely reason for a child under five to be hospitalised with croup[332] – and croup isn't the only serious illness they cause.

Parainfluenza virus infections have a nasty habit of spreading down the respiratory tract to the bronchioles, which connect the lungs to the trachea. Inflammation of the bronchioles, called bronchiolitis, can block the movement of air in and out of the lungs. That's as serious as it sounds. In some cases, parainfluenza viruses get into the lungs themselves and cause pneumonia, which interferes with the uptake of oxygen[333].

The more serious effects of parainfluenza viruses usually happen in young children, the elderly or people whose immune systems are weakened such as by AIDS or malnutrition. They rarely cause anything worse than a cold in a healthy adult but, as Robert Chanock went on to show, they have a relative that's even more likely to cause a dangerous bronchiolitis.

A chimpanzee called Sue

The influenza viruses, the adenoviruses, the rhinoviruses and the parainfluenza viruses were all discovered by researchers proactively looking for people with the sort of respiratory illnesses that cold viruses cause. The next cold virus was first discovered in the throat of a chimpanzee called Sue[334]. She was one of 20 chimpanzees kept for research purposes at the institute that Hilleman and Werner were working for when they discovered the first: the Walter Reed Army Institute of

Research, as the Army Medical Service Graduate School had recently been renamed.

In 1955, something was making the Walter Reed chimpanzees cough and sneeze like a class of schoolchildren during a seasonal influenza outbreak. In another place or at another time, an outbreak among research animals would have been regarded as an inconvenience to be waited out but in 1955, cell culture had made every unexplained malady into an opportunity to isolate a new virus. This particular virus refused to infect any animal other than chimpanzees except, possibly, for one of the researchers who came down with a cold. His colleagues took a sample of his blood and sure enough, it contained antibodies that inactivated the virus they'd isolated from Sue.

A quarter of a century earlier, Alphonse Dochez had shown that human cold viruses would infect chimpanzees but no other animals*. Here was a virus that must have been brought into the chimpanzee house by a human, had infected the chimpanzees and presumably infected another human. It looked remarkably like the cold virus as described by Dochez and every other cold virus that refused to infect any animals other than humans and chimpanzees.

The Walter Reed researchers were confident they had isolated a new virus, but they couldn't be sure how important it was. They published a description of what they called 'chimpanzee coryza virus' with the appropriate qualifications[335] and moved on to other things.

The naming of RSV

Meanwhile, Robert Chanock had left Sabin's well-funded but overbearing mentorship in Cincinnati for a post at Johns Hopkins University in Baltimore. He was still looking for the viruses that were hospitalising children with respiratory problems but now, he was looking further down the respiratory tract. He wanted to know if it was only parainfluenza virus that was hospitalising children with inflammation of the bronchi, bronchioles and lungs or if there was another virus involved.

He did indeed find another virus which he called the 'Long-Snyder' virus[336] after the surnames of the two individuals he isolated it from.

---

* Chapter 1 described Dochez's chimpanzee experiments.

Research ethics had still not got around to the idea of anonymising research subjects. Chanock ran the battery of chemical tests that virologists used to characterise new viruses and found the Long-Snyder virus was very different to any known human virus. However, it had exactly the same properties as the virus that had been isolated from Sue the chimpanzee.

Chanock applied serum containing antibodies that neutralised chimpanzee coryza virus to his Long-Snyder virus and sure enough, the serum neutralised it. The virus that had caused a cold in captive chimpanzees in Washington was the same virus that was hospitalising children in Baltimore.

Now Chanock knew what he was looking for, he started finding a lot more of it. In the winter of 1956 to 1957, he took throat swabs from children with otherwise unexplained pneumonia or bronchiolitis from several Baltimore hospitals. Again and again, he found the Long-Snyder virus was causing their illness.

When Chanock used monkey kidney cells to culture the new virus, it broke down the membranes between the cells so that they formed a single mass of material from several fused cells. Such a mass is called a syncytium, which prompted Chanock to suggest a new name: the respiratory syncytial virus[337]. The name remains in use, usually abbreviated to RSV, which is somewhat misleading because it was later found that unlike some of the other cold viruses, RSV only rarely causes syncytia to form in someone's respiratory tract[338]. The syncytia that Chanock observed were an artefact of his cell culture system.

Nevertheless, the name has stuck. I warned you about viral etymology.

In less than five years and with a break for military service, Chanock had largely filled the knowledge gap that had bothered him since he treated that little girl's croup. Between them, parainfluenza virus and RSV account for the majority of children being hospitalised with croup and bronchiolitis[339], and a substantial number of pneumonia cases.

They also cause a lot of straightforward colds.

It was a resounding scientific success and a major step toward better treatments for those conditions. Chanock was about to discover that a medical scientist's successes in medical science don't always carry much weight in the realm of office politics.

He became aware that one of his colleagues was publishing fraudulent research and reported him to the dean of the Johns Hopkins School of Medicine. The dean responded in the time-honoured fashion of a senior administrator faced with a whistle-blower: he demanded Chanock's resignation[340].

More than one promising scientific career has ended when a burst of integrity encountered administrative defensiveness. Chanock received a lifeline in the form of a post at the US government's National Institute for Allergy and Infectious Diseases, which was and remains one of the world's leading medical research centres. He spent the rest of his career there.

The man he reported remained unscathed by the incident until, in Chanock's words, he 'was caught with his hand in the cookie jar about fifteen years later and was summarily dismissed.'[341]

The respiratory syncytial virus: not just a cold

The connections between parainfluenza virus and RSV went beyond both having been discovered by Chanock. They are very similar biologically and for some time, virologists placed them in the same family. In keeping with virological tradition, that family was named with a jaw-breaking Greco-Latin polysyllable: the paramyxovirus family, which also contains the measles and mumps viruses.

Since then, virologists have recognised that RSV is less similar to parainfluenza virus than was first thought and placed it in a newly coined family called the pneumoviruses[342]. 'Pneumo' is a Greek word that could translate as either 'lung' or 'breath', the latter being more apposite because RSV is transmitted through the breath but only rarely infects the lungs.

As with parainfluenza virus, Chanock first identified RSV from children it had hospitalised, and as with parainfluenza virus, only a small proportion of people it infects end up in hospital. The vast majority simply get a cold. In adults, RSV accounts for about the same proportion of colds as parainfluenza virus: about one in 20.

Once researchers knew to look for RSV in children, however, they found a very different picture. The two subtypes of RSV amount to one of the most serious diseases of early childhood. Among any ten children, two will suffer an RSV infection in their lower respiratory tract by the time they're one year old[343]. The lower it gets, the worse it is for the child.

The most frequent consequence is bronchiolitis, which is how it first caught Chanock's attention, although RSV can cause bronchitis and pneumonia as well. It kills over 100,000 children every year[344], which is a very small percentage of the children who catch RSV, but those fatal cases typically have two things in common.

First, more than 19 out of 20 of those RSV deaths happen in the poorer parts of the world[345]. A child with bronchiolitis who gets to a well-equipped hospital is in for a miserable experience, but they're probably going to survive it. Paediatric units are very good at keeping air going in and out of tiny lungs, whatever inflammation they may have to get it past. Where no such unit is available, the outlook for a child with bronchiolitis is a lot bleaker.

Second, most of those deaths are in babies under six months old. Younger and smaller babies have smaller respiratory tracts and the narrower a bronchiole is to start with, the more prone it is to be completely blocked by RSV-induced inflammation.

There's another reason why younger babies are more likely to have a more serious bout: they have no immune memory until their first RSV infection. They respond to that infection by flooding the bloodstream with antibodies which protect them from subsequent RSV infections – albeit up to a point. Those antibodies protect the lower respiratory tract, which prevents pneumonia and RSV's signature illness of bronchiolitis, for decades after the infection. On the other hand, immune memory doesn't protect the upper respiratory tract for long. Consequently, we get infected with RSV every few years, but it rarely causes anything worse than a cold[346]. That immunity tends to break down later in life, which leaves elderly people at risk of serious complications from RSV infection, even if they've fought it off dozens of times in the preceding decades.

Chanock's discovery of RSV led to the discovery of the serious risk it poses to babies, which inevitably led to the question of what might be done about it. At first glance, the solution seemed obvious: if immune memory to RSV protects the lower respiratory tract, then why not give a baby that immune memory before they're infected?

It sounds simple, but it's taken five decades to achieve.

The Lot-100 debacle

By the time Chanock joined the NIAID in 1957, vaccination was a tried and tested way to induce immune memory and, thanks to the recent breakthroughs in cell culture, American researchers were developing vaccines against viruses that had recently seemed impossible to prevent. The first major success came when the first polio vaccine proved successful in 1955. From New York to Nebraska, the triumph was celebrated with ringing church bells and wailing fire sirens[347].

It created a heady atmosphere of vaccine optimism. If these new cell culture techniques paved the way to a polio vaccine, then why not all the other viruses that had so recently seemed so intractable.

Why not RSV?

The principle behind vaccination is simple: isolate the microbe you want to vaccinate against, modify it in a way that prevents it from causing disease but can still induce immune memory and then inject it into the person in whom you want to induce immune memory. The practice is considerably more complicated than the principle because there are so many different ways to modify a microbe and even today, it's impossible to predict which way induces immune memory that protects against the unmodified microbe without trying it.

Chanock opted for the same approach that had worked for that polio vaccine, which was to produce an 'inactivated' vaccine. He harvested cultured virus in the form it takes outside living cells and disrupted the capsid's chemical structure with a chemical called formalin. The capsid proteins would all be there for the immune system to process and produce antibodies against, but they'd be too disrupted to get the virus inside the cells of anyone it was injected into. If the virus couldn't get into the cells, it couldn't replicate and it couldn't do any harm.

At least, that was the idea.

The only way to find out if it worked was to inject it into babies. It had to be babies because Chanock wanted to protect against the most dangerous RSV infections, and it's in babies that those infections happen.

The trial of the vaccine he called Lot-100 would become one of vaccine research's most important cautionary tales.

When Lot-100 got to the point of being ready to test, there was a tried and tested approach to finding children to test vaccines and treatments on. Unfortunately, it was an approach to which a modern ethics committee would respond with reams of notes that could be summarised as 'absolutely not'.

From a purely scientific perspective, institutionalised children were the ideal subjects. They lived in the same buildings, they ate the same food and they were all exposed to the same microbes at the same time. Dormitories full of children made for a perfect environment for spreading airborne viruses like RSV, so if Chanock's team vaccinated some children and left others unvaccinated, they could tell whether or not the vaccine protected children in one winter cold season. A children's home was the next best thing to a laboratory animal house with the added advantage that a researcher didn't need to ask the permission of every child's parents. A blanket agreement from the home's management gave them access to every child.

The year of the Lot-100 trial, 1966, was also the year in which a seminal paper would ask the question of whether this sort of thing was ethically acceptable[348]. That paper would initiate a debate which, after a great deal of soul searching, tooth gnashing and – inevitably – endless committee meetings, would lead to general agreement that it was unethical to test new medications in children's institutions and establish ethics committees to ensure that they were not.

In 1966, all of that lay in the future. Institutions were still the preferred venue for vaccine trials and if Chanock and his colleagues picked what would become one of the most notorious children's institutions in America, it was probably because of its proximity rather than because of that notoriety.

Junior Village in Washington DC was within easy striking distance of the NIAID and it was one of the largest children's homes in the USA – although it probably shouldn't have been. It was originally designed to hold 320 children but a recent crackdown on supposed welfare fraud, aimed mostly at the black districts of Washington DC, had led to so many children being separated from their parents that Junior Village now held around 900 children aged between six months and 18 years. The sudden influx had so exceeded the capacity of Junior Village's buildings that many children slept in hastily acquired army tents erected in a compound sited between a refuse dump and a sewage treatment plant[349].

Despite the overcrowding and the insalubrious location, Junior Village was widely touted as a charitable triumph. Regular visitors included Martin Luther King and Jacqueline Kennedy, and Junior Village children were regularly invited to Christmas parties at the White House.

Years later, a darker truth emerged: when the cameras weren't watching, life in Junior Village was brutal. Younger children were subject to constant physical and sexual abuse from counsellors and older children alike, and there was never enough food to go around. One inmate – or, perhaps, survivor – later recalled that when he was transferred to a different orphanage at 12 years old, 'I had to ask what boiled eggs were since I'd never seen them before.'[350]

It's unlikely that King and Kennedy would have been so enthusiastic about the place if they'd known what was going on; if nothing else, it was a brewing scandal. Chanock and Albert Kapikian, Chanock's colleague who led most of the Junior Village work[351], can't have been completely oblivious to Junior Village's problems, but they may have felt that withdrawing was likely to make matters worse rather than better. The NIAID projects came with a paediatrician who was able to examine and treat children who fell ill, giving them much better medical care than those in most institutions. It's easy to condemn the ethics of working with Junior Village children today, but at the time, walking away and leaving them to it would not have been an ethically uncomplicated option.

The Lot-100 trial got off to a good start. Of 27 children vaccinated, 26 produced antibodies against RSV.

So far so good.

Then, in December 1966, RSV arrived in Junior Village. Amid the usual barrage of sniffling among the older children whose immunity protected them from anything worse than a cold, Chanock's team were monitoring their vaccinees and another 47 children under five years old who were about to experience their first RSV infection. Among those 47 unvaccinated children, four developed serious enough complications to land them in hospital[352]. That's a high proportion of hospitalisations[353], hinting at the benefit of having a paediatrician on site. It's hard to imagine that a management regime that didn't feed the children properly would have been so diligent about making sure they were hospitalised when they needed to be.

If the vaccine worked, a far lower proportion of vaccinated children would need hospitalisation. Instead, more than two-thirds of the vaccinated children developed serious complications and two died[354]. Instead of protecting children from RSV, Lot-100 made it far worse.

There were several consequences to the Lot-100 debacle. The first was that Lot-100 was consigned to the dustbin of medical history, along with any other attempt to make an inactivated RSV vaccine. A wider-reaching consequence was the discovery of what's now called vaccine-associated disease enhancement.

It hadn't occurred to Chanock or his colleagues that Lot-100, or indeed any other vaccine, might make a disease worse rather than better. Since then, every vaccinologist testing a new vaccine has been painfully aware that they need to consider the possibility.

The debacle created a conundrum for RSV research that would last for decades. On one hand, RSV kills a lot of babies whom the right vaccine might protect. On the other, the only way to find out if an experimental vaccine was that right vaccine was to test it on babies which, after the Lot-100 debacle, was a risk that nobody was willing to take.

Rethinking RSV immunisation

Doctors and scientists were forced to take a different approach. The antibodies dissolved in the bloodstream are a key element of immune memory and if it was too dangerous to try to vaccinate a baby into producing their own antibodies, an alternate approach is to inject antibodies directly into a baby's bloodstream.

If that sounds straightforward, it isn't.

One cannot simply take serum from an immune person and inject it into someone else. The antibodies used as a treatment are produced in carefully nurtured cultures of identical cells and subjected to a complicated process of purification and quality control. The process is complicated enough to make monoclonal antibodies horribly expensive and they can only be given in a well-equipped hospital.

In high-income countries with well-funded healthcare systems, monoclonal antibodies are available to treat serious RSV infections. Moreover, they're often given pre-emptively, to protect babies born prematurely who are at particularly high risk of RSV bronchiolitis.

However, antibody treatments are rarely available in low-income countries, which is one reason why such countries lose far more babies to RSV.

There is another way to get antibodies into newborn babies that doesn't involve injecting anything. During pregnancy, some of the mother's antibodies cross the placenta to the foetus so when the baby is born, he has a cocktail of 'maternal antibodies' in his bloodstream. They'll only protect him through his first few months because until his own immune memory is induced, he has no way to replenish them himself, but they do give him some protection during those vital first months.

With RSV, that protection depends on how recently RSV last infected his mother. Every RSV infection gives the immune system a kick up the backside, to which it responds by raising the concentration of antibodies in the bloodstream. In the months and years following that infection, the antibody concentration gradually falls. That means that a baby whose mother recovered from RSV recently will receive a higher concentration of antibodies that will last longer than a baby whose mother hasn't been infected for some time, who may not receive enough to protect him for the critical first six months when RSV is most dangerous.

After the Lot-100 debacle, some vaccinologists wondered if maternal antibodies might be the safe solution they're looking for. Lot-100, it appeared, had directed the immune response in a dangerous direction but that was because it had been the first exposure to RSV for the children it was given to[355]. A woman old enough to be pregnant would already have been infected several times, so she'd already have a well-established immune memory to RSV.

A vaccine wouldn't shape her immune response, the reasoning went, but it might raise the concentration of antibodies transferred to her foetus and extend the length for which they protect the baby after he's born.

'Maternal vaccination', as vaccinating a pregnant woman to protect her child is called, is widely used against whooping cough and influenza. Perhaps it could work against RSV.

The theory was sound. The practice proved more difficult. In the decades following the Lot-100 trial, several vaccines were tried and, while none did any harm, none did much good. Success remained elusive until 2023 when, after decades of one candidate vaccine after another being discarded as ineffective, not one but three different vaccines were reported to be effective against RSV.

Only one of the three reports described successful maternal vaccination. The vaccine in question was made by the pharmaceutical firm Pfizer under the brand name Abrysvo and was given between the 24th and 36th weeks of pregnancy. It cut the risk of serious lower respiratory tract infection, including RSV's signature nastiness of bronchiolitis, by more than half. It was even more effective against the severe cases likely to land a baby in hospital or worse, which it cut by more than two-thirds[356]. It would have been nice to have completely prevented those serious cases but it's considerably better than anything else that's available. Within a year of the report's publication, it was rolled out in the USA[357], Britain[358] and approved in the European Union[359] which enables national healthcare systems to offer it.

All three vaccines proved effective in the other population at high risk of serious RSV infection: older adults. One was GSK's Arexvy[360] which, like Abrysvo[361], prevented around two-thirds of lower respiratory tract infections in adults over 60 years old. If those two vaccines have similar effects, it's probably because they are made in much the same way: by isolating a protein from RSV's capsid that's key to getting the virus inside the cells it infects.

The third vaccine is made in a different way; mResvia, made by Moderna, uses the same mRNA technology that's become familiar through the most widely used COVID-19 vaccines. Instead of simply injecting the viral protein, it mimics the process by which RSV produces that protein. The eponymous mRNA is a chemical very similar to that which encodes RSV's genome and like that genome, it's inserted into cells which it co-opts to produce the protein.

In the past, persuading a cell to synthesise a viral protein has usually proved more effective than simply injecting the protein, probably because it triggers the immune processes that evolved to fight off viruses. However, mResvia does not appear to be markedly better than either of the two protein-based vaccines. The reported efficacy was marginally better, in that Abrysvo's and Arexvy's prevention of two-thirds of lower respiratory tract infections was at the lower end of the range of mResvia-induced protection[362] rather than in the middle of it. Nevertheless, the three vaccines were in the same ballpark and without a head-to-head comparison in a single trial, we can say that mResvia *looks* a bit better but we'll have to wait to see how all three shape up.

The reason we have to wait is that, at the time of writing, RSV vaccination is still in its very early days. While all the RSV clinical trials involved thousands of individuals who were carefully monitored, that's far fewer than the millions who will receive the vaccines in the coming years. Despite that uncertainty, there is good reason to believe that a baby born today has a far better chance of weathering his first RSV infection than a baby born as recently as 2022, before any RSV vaccines were available – if, that is, the baby happens to be born in a country in which RSV vaccines are available. At the time of writing, we can only hope that RSV vaccines will soon become available in the low-income countries where RSV kills the most babies.

It would be fair to say that the Lot-100 debacle was not the highlight of Robert Chanock's career, but it would be unfair to define his legacy by it. He was promoted to chief of the NIAID's Laboratory of Infectious Diseases in 1968 and he held the post until he retired in 2002[363]. He was heavily involved in his department's many achievements which included one of the first antibody treatments for RSV and the discovery of so many disease-causing microbes that Anthony Fauci, who became Chanock's direct superior when he was appointed to direct the NIAID in 1984, said that when he was first learning about infectious diseases, 'Bob's papers ... popped up everywhere. The name "Chanock" seemed synonymous with disease discovery.'[364]

Back in the late 1950s, Chanock's discovery of the parainfluenza viruses and RSV took cold virus research a solid step beyond the earlier discoveries of adenoviruses and rhinoviruses but, at the time, that was as far as the cold virus discoveries were likely to go. Those four virus groups were the first to be discovered because they were the ones that could be isolated with the cell culture techniques then available.

Virologists were swabbing cold-afflicted throats as enthusiastically as ever and in many cases, the viruses causing those colds would not replicate in their cell cultures. Once again, cold virus research was at an impasse.

This time, it wouldn't take two decades to escape it.

Chapter 9

Enter the coronaviruses

Whenever a new decade arrives, it's always tempting to look for some seismic change in the arbitrary numbers assigned by the Gregorian calendar. Such changes are not always there to be found but in 1960, they were plain to see.

British prime minister Harold Macmillan welcomed the new decade by speaking of the 'wind of change' blowing across Africa[365], publicly acknowledging the imminent demise of the British empire. In Liverpool, an obscure skiffle band called Johnny and the Moondogs renamed themselves The Beatles. Across the Atlantic, the US government began deploying troops into the escalating conflict in Vietnam a few months before John F. Kennedy became the youngest man ever elected to the presidency.

In the world of cold research, changes would be slower to arrive. Some of those changes stemmed from advances in cell culture. Virologists and cell biologists continued to expand the range of cells that they could keep alive which enabled them to culture a broader range of viruses.

Yet there was a much more significant change on the way. After more than half a century of characterising viruses by their effects on animals, on humans and on cell culture, the electron microscope enabled virologists to see viruses directly for the first time.

Images of viral capsids – the protein structure that coats the genome between cells – revealed that viruses with similar characteristics also had very similar capsids. By the end of the 1960s, the capsid's size and shape were a key part of how viruses were described and classified.

Virologists working on colds entered the new decade knowing that the cold viruses they had yet to isolate were the more difficult ones that had refused to yield their secrets to the cell cultures of the 1950s.

It took novel approaches to both cell culture and electron microscopy and several years of frustration to isolate the first coronavirus.

Today, the word 'coronavirus' is often used synonymously with SARS-CoV-2. However, virologists were studying SARS-CoV-2's relatives long before SARS-CoV-2 emerged in 2020 and for a long time, they didn't think the coronaviruses were very important at all.

The boarding school virus

The British boarding school is the core of a whole subgenre of literature. A shelf of such tomes might place the hilarious hijinks of the St Trinian's cartoon strip next to the bleak reminiscences of Robert Graves and Roald Dahl. The common themes are casual flogging by masters, tyrannical bullying, terrible food and cold, damp and crowded dormitories. The scions of the British upper classes were formed in an atmosphere that could be as miserable as Junior Village and was just as conducive to transmitting colds.

That made boarding schools very interesting to CCU scientists, who had an ongoing collaboration with the doctor who oversaw a boys' boarding school that was never named in their reports. In the winter of 1960 to 1961, they swabbed every virus they could out of sore throats and stuffed noses. Several turned out to be rhinoviruses that replicated in the cells they were able to culture, but others proved more difficult. Those difficult viruses would cause colds when they were inoculated up the noses of CCU volunteers, but they refused to infect any of the cultured cells in the CCU laboratory[366].

The effort was led by David Tyrrell, whose work had been instrumental in isolating rhinoviruses a few years earlier*. He continued to lead it when he was promoted to CCU director in 1962[367], when Andrewes retired[368]. Tyrrell had a pretty good idea of the problem: he still couldn't culture the cells from the respiratory tract lining, which are the cells that a cold virus would naturally infect. Sure, the HeLa cells and kidney cells they were using had *some* similarities with the cells lining the nose and the throat, which was why *some* cold viruses

---

* Chapter 7 describes how the CCU scientists co-discovered what they called the ECHO viruses before they were renamed the rhinoviruses.

could infect them, but there were also a lot of differences. It was hardly surprising that a lot of cold viruses couldn't.

Then Andrewes was invited to Sweden to receive an honorary degree from the University of Lund. While he was there, he met an ear, nose and throat surgeon called Bertil Hoorn who was well on the way to being able to culture cells from the throat.

Like Tyrrell, Hoorn had tried to culture the cells that line the throat and found they promptly died. He could tell by watching the cilia which, when those cells were still in someone's throat, had been keeping the mucociliary escalator moving. As long as the cells were alive, the cilia keep waving but, as soon as Hoorn put a sample of those cells in culture, the cilia stopped moving as the cells died.

Up to that point, scientists setting up cell cultures had tried to separate the different cell types and culture them separately. Such an approach had worked for George Gey when he established the HeLa cell line and for numerous other scientists since but, when it came to throat cells, Hoorn wondered if that might be where everyone was going wrong.

He tried cutting sections of the respiratory tract of rabbits, containing not only the ciliated cells of the tract lining but also the layers of tissue underneath them, and placing them in culture dishes. This time, when he viewed them under a microscope, Hoorn saw the cilia waving back at him. They kept waving for more than a month.

He tried the same approach on human cells, which he sampled from the respiratory tracts of aborted foetuses. It worked just as well.

Hoorn had been right: those ciliated cells did indeed need other types of cells to keep them alive.

Now he could keep those ciliated cells alive, Hoorn's next move was to see if viruses would replicate in his cultured epithelium cells. He inoculated his cultures with adenoviruses or influenza viruses and watched the cilia stop waving as the viruses co-opted the cells' metabolism into making new viruses, killing the cells in the process[369].

When Andrewes saw that Hoorn had not only succeeded in culturing cells from the respiratory tract epithelium but also succeeded in using them to culture cold viruses, he invited him to the CCU[370]. Perhaps these 'organ cultures', as Hoorn called them, would get those intransigent boarding school viruses out of the volunteers' huts and into the laboratory.

Hoorn puttered his way to Salisbury in his Citroën 2CV and showed Tyrrell how to establish organ cultures from the respiratory tracts of aborted foetuses. Tyrrell inoculated those human respiratory tract cells with virus B814, one of the recalcitrant boarding school viruses. Over the next few days, he watched the cilia of the inoculated cells fall still. That was a good sign that the virus had infected and killed those cells but to prove it, he needed to confirm that B814 was replicating in them. He filtered some of the nutrient medium and dripped it up the noses of twelve volunteers. Ten of them developed colds, which was an unusually high success rate for the CCU's inoculations. He passed B814 between volunteers and cell cultures several times before he was convinced that he had indeed succeeded in culturing B814.

He then ran it through the standard series of chemical tests and checked that it wasn't inactivated by antibodies to any known virus. By 1965, he'd established that B814 had such a unique set of characteristics that it was not only a new type of virus but an example of a hitherto undiscovered virus family.

Later in his career, Tyrrell would celebrate one of his many successes by loudly singing 'Praise, My Soul, the King of Heaven' while mowing his lawn[371]. Whether he did something similar to celebrate his taming of B814 is unrecorded.

So far, he'd characterised B814 using well-established techniques like serotyping but then he wondered if he might be able to go a step further. After more than half a century of virology, it was at last possible for a virologist to see a virus.

Tyrrell wanted to put B814 under an electron microscope.

To visualise the invisible

Since medical scientists first realised that microbes caused disease back in the late nineteenth century, the light microscope has been one of their most important tools. Generations of medical microbiologists have magnified their subjects by up to a thousand times – but no more. That's a limit set by the physics of visible light that no lensmaker can ever improve on.

It's enough to see how human cells organise themselves into tissues and enough to see some detail of those cells' structures. Bacterial cells are much smaller, but a thousandfold magnification still shows their shape in enough detail to identify broad groups of bacteria if not always the species.

Viruses, however, are a different matter. Their capsids are so small that they need to be measured in nanometres, each of which is a millionth of a millimetre. On a more human scale, a nanometre is slightly less than a fingernail grows in a second[372]. The figure on page 154 shows some of the cold virus capsids to scale, the largest of which is the parainfluenza virus at 150–200 nanometres across[373], or between three-and-a-quarter and four-and-a-half minutes of fingernail growth. The roughly 30-nanometre rhinovirus is less than 30 seconds of fingernail growth[374].

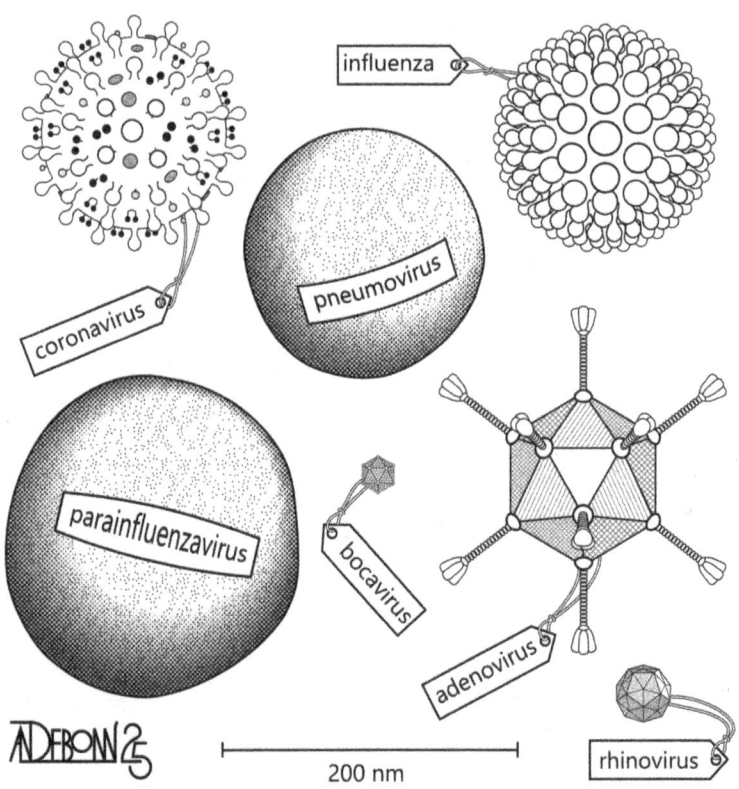

The capsids of the major cold viruses described in this section, derived from electron microscope and electron micrograph images. The scale bar represents 200 nanometres or 1/5,000 millimetre. The pneumovirus family includes both the respiratory syncytial virus and the metapneumovirus, which have very similar capsids.

It's not possible to show a human cell on that scale. If that parainfluenza virus were magnified up to the size of a marble, one of the lymphocyte cells that chases it out of the respiratory tract would be as big as a basketball.

Even the best light microscopes, with their thousandfold magnification, aren't powerful enough to explore the nanometre scale at which viruses operate. That's why Hoorn and Tyrrell had to characterise them by inference. They couldn't see the virus directly but they could see if it caused a cold in a CCU volunteer or stilled the cilia of a cultured cell.

An electron microscope works in a fundamentally different way to a light microscope. It fires a beam of electrons at a sample and builds a picture based on how they interact with it. Those electrons interact with samples that are far too small to interact with the visible light used by light microscopes, enabling an electron microscope to give up to 100 times the magnification of a light microscope. That's enough to see the capsid in detail.

When Tyrrell managed to culture B814 in the mid-1960s, it was more than 30 years since Ernst Ruska produced the first electron beam images. Ruska picked a particularly inauspicious time and place to demonstrate a new concept: the Technical University of Berlin in 1928[375]. The following year's economic crash dried his funding to a trickle and although he proved his concept by producing the first electron microscope images in 1933, he had to look for a new job.

Ruska spent a few years making ends meet wherever he could until the engineering firm Siemens-Schuckert*, where someone was evidently impressed by his images, took him on to develop it further. Ruska had always intended his invention to be used in the medical sciences and virology was an obvious application because it was, by definition, the study of entities too small to be seen any other way. By 1939, Ruska could achieve 25,000 times magnification and proved it with a set of images of a plant virus capsid[376].

At a different time or a different place, the first image of a virus would have been a pivotal discovery in virology. However, Ruska published his image in a German-language journal while the world

---

* Now part of the conglomerate Siemens AG.

was still coming to terms with the German invasion of Czechoslovakia and wondering who was going to be next. Even among the virologists of the day, few noticed that someone had finally captured a picture of the mysterious entity they were studying.

Ruska's paymasters at Siemens-Schuckert were less interested in the major breakthrough that those images were in themselves than in the potential they demonstrated. The Nazi government was setting the priorities and virology was not among them. The money was in weapons manufacture, which electron microscopy could improve by imaging the edges of cutting tools and the metal that they cut in minute detail. The electron microscope became a military technology and while Ruska had already published too much to make it a complete secret, it was not publicised any further. Until the end of the Second World War, very few people outside Germany had any idea that such a thing as an electron microscope existed.

To tens of millions of people, the collapse of Nazi Germany in 1945 was a release, a reprieve and a cause for celebration. To Siemens-Schuckert, it was the loss of their main customer – but it opened new opportunities. A company that had operated workshops through the Auschwitz extermination camp[377] pivoted to being a major exporter of technology to the countries that occupied West Germany. The electron microscope became a flagship product and many of Siemens-Schuckert's customers wanted it for the medical applications that Ruska had originally envisaged.

It would win Ruska the Nobel Prize in physics, albeit not until 1986, but the electron microscope was still several innovations away from the toolkit of human virology.

The problem was that with such a high level of magnification, it's only possible to look at one tiny part of the sample at a time. Ruska had produced those images of a plant virus because it was easy to work with, enabling him to produce a very high concentration of capsids in the sample. With human viruses, the vagaries of cell culture were such that the samples contained much lower densities.

To get a sense of their problem, open Google Earth and try looking for ships or aircraft in the Pacific Ocean. They're out there, wherever one of them happened to be passing below Google's satellite when it took the picture. However, you can't see one until you've zoomed in to such a tiny piece of the vast Pacific that the chances of finding a ship or a plane where you happen to look are so low that you may as well forget it. As with ships and planes in the Pacific, so with capsids in a sample: there's vastly more space between them than occupied by them.

Before virologists could get much value out of an electron microscope, someone was going to have to come up with a better way to prepare virus samples.

The defining crown

In a field dominated by men who spent their early adulthood accumulating degrees, a woman who finished her education when she left school at 16 was unusual. University wasn't seen as an option in the Glasgow tenement where June Hart was brought up. She worked as a junior laboratory technician for a few years and then married and moved to Toronto with her husband, a Venezuelan artist called Enriques Almeida. The marriage didn't last but the move set her on a career path that would make June Almeida – the name she kept for the rest of her life – one of the best-known names in her field.

That path started with a job at the Ontario Cancer Institute where she was introduced to the electron microscope. The opportunity to explore the minuscule fascinated her and she showed an aptitude for improving on established techniques. She developed such a reputation that in 1964, London's St Thomas's Hospital headhunted her back to Britain.

Her new post led her to apply electron microscopy to virology and she cemented her reputation with the first images of the capsid of the rubella virus[378] which causes German measles. It established her as the person to call when a virologist had a problem that might be solved with an electron microscope.

One such virologist was David Tyrrell, who wanted Almeida to show him what his B814 virus looked like.

Almeida was happy to join the hunt for cold viruses and assured Tyrrell that she'd be able to image B814. Tyrrell was dubious. His years of struggle with B814 had persuaded him that it didn't give up its secrets without a fight and he was painfully aware that for any scientist, getting carried away with the shiny new technique you've invented is an occupational hazard. It wouldn't be the first time Tyrrell had known someone to promise results they couldn't deliver. He'd probably done it himself once or twice.

Almeida had not earned her reputation by overpromising. Tyrrell later wrote that her results 'exceeded all our hopes'[379]. Her images of the elusive B814 were far clearer than Tyrrell had dared hope for.

For Almeida, those images jogged a memory. She'd seen two very similar viruses before. One came from a liver infection in a mouse and one from the respiratory tract of a bird. She'd written a paper stating that the two viruses belonged to a new family, but it had been rejected. The reviewers did not agree that they were new viruses but believed they were poor-quality images of influenza viruses.

Almeida showed those images to Tyrrell, who agreed she'd been right and the reviewers had been wrong. Almeida's animal viruses and his B814 bore a passing resemblance to influenza viruses but they were much more similar to each other. Tyrrell was as sure as Almeida that they belonged to a different family.

All three viruses sported what looked like a halo, or perhaps a crown, but Tyrrell and Almeida could hardly call them haloviruses or crownviruses. In the world of 1960s virology, naming viruses using plain English simply wasn't done.

Despite the liberal use of Greek and Latin in virus nomenclature, very few 1960s virologists spoke either language and neither Tyrrell nor Almeida was among them. They reached for a dictionary and found that the Latin for 'crown' was 'corona' so they called their new family the coronaviruses[380].

Almeida's images of B814 were published in 1967[381], seven years after it made some young boarder's winter even more miserable. Getting from a schoolboy's throat to an electron microscopy image had needed Bertil Hoorn's tissue cultures, June Almeida's magic with the electron microscope, David Tyrrell's green fingers in the tissue culture laboratory and a lot of CCU volunteers who put up with deliberately induced colds.

And that was only the beginning of the coronavirus story.

The American isolations

In 1962, Dorothy Hamre of the University of Chicago isolated a cold virus by infecting a culture of kidney cells. Five years before Almeida and Tyrrell identified the coronaviruses as a distinct family, she had no way to know her isolate was a member of the same family as the B814 then being passed from volunteer to volunteer at the CCU. Hamre put her virus through the usual battery of chemical treatments and serum that inhibited known viruses and concluded that the virus she called 229E was something new[382].

Meanwhile, at the National Institute of Allergy and Infectious Diseases, Kenneth McIntosh was looking for cold viruses under the direction of Robert Chanock. He took samples from any of his colleagues who had a cold, although whether Chanock himself contributed any samples is not recorded. Like Hamre, McIntosh tried to infect whatever cells he could culture, but he needed to use Hoorn's organ culture technique to isolate what he called virus OC43. He applied the same battery of tests as Hamre and similarly concluded that he'd isolated a new virus[383].

Hamre and McIntosh quickly established that antibodies that neutralised either one of their viruses did not neutralise the other, showing that 229E and OC43 were different viruses. Hamre sent a sample of 299E to the CCU, where it went straight up the noses of volunteers and caused colds that tended to be nastier than those caused by rhinoviruses but similar to those caused by B814[384]. Sure enough, when Almeida worked her magic on 229E, she saw the crown that identified it as a coronavirus[385].

Meanwhile, Hamre was finding that 229E was rife among the University of Chicago's medical students[386]. It was a prolific cause of the common cold as, a study conducted between 1963 and 1970 would show, was McIntosh's OC43. Between them, 229E and OC43 caused between one in ten and one in fifteen spring and winter colds in American adults[387] and many later studies done at different times and different places have shown broadly similar proportions.

Unfortunately, B814 appears to have been lost soon after Tyrrell and Almeida's seminal paper[388] and there's no record of anyone testing whether antibodies that inactivated B814 also inactivated either Hamre's 229E or McIntosh's OC43. It's a peculiar omission, especially as the CCU must have stored antibodies against B814, but if anybody tested them against the other two coronaviruses, they did not publish the results.

Given that McIntosh had needed organ culture to isolate OC43 while Hamre used conventional cell culture techniques to isolate 229E, B814 and OC43 were probably different isolates of the same virus. In the absence of a definitive test, 'likely' is as far as the reasoning can take us and B814's loss left 229E and OC43 as the defining isolates of the first two human coronaviruses to be recognised.

By then, Tyrrell and Almeida had both moved on to other viruses. Tyrrell briefly returned to the coronaviruses in the late 1980s, when he looked into how long immunity to them lasted. He infected a group of volunteers with 229E and found he could infect the same volunteers with the same virus a year later. His experiment showed that as with many of the cold viruses, immunity to 229E was short-lived[389]. We can expect repeated infections with 229E and indeed the other cold-causing coronaviruses throughout our lives[390].

Tyrrell's last coronavirus study was published in 1990, the year in which the CCU was closed. He moved to the Centre for Applied Microbiology & Research* a few miles away, where he was involved in several research projects including pioneering research on viruses associated with chronic fatigue syndrome[391].

Almeida's work on coronaviruses ended with those seminal images of B814 and 229E. Like most scientists, she never became famous in her lifetime although in 2020, the COVID-19 pandemic sparked a brief flurry of interest in the woman whose images were so central to the discovery of the coronaviruses. One commentator went as far as to acclaim her the 'virus queen'[392], as if she'd had a crown bestowed upon her by a virus instead of naming a virus for its own crown.

As hagiographic as some of the articles about her were, few showed much interest in her later work, which was at least as important to modern virology. She developed a technique called immune electron microscopy that completely bypassed the need to get a virus replicating in a cultured cells and enabled her to visualise hitherto unknown viruses sampled directly from a clinical sample.

Virologists from all over the world travelled to London to learn how to identify a virus without having to contend with the vagaries of cell culture. Among them were Robert Chanock and the man who led the NIAID's work at Junior Village†, Albert Kapikian[393]. Chanock and

---

* Operated by the government's Department of Health and now part of the UK Health Security Agency.

† Chapter 8 describes the disastrous respiratory syncytial virus vaccine trial that Kapikian oversaw at Junior Village.

Kapikian credited what Almeida taught them with their subsequent discovery of the norovirus[394], notorious as the winter vomiting bug.

Moreover, Almeida played a key role in introducing electron microscopy into virology. Viruses were no longer obscure entities that could only be discerned by their effects on cells. She was at the forefront of developing the techniques that revealed the many different shapes and sizes of viral capsids.

She retired – or possibly abdicated – from science in 1985 and started a whole new life teaching yoga, restoring chinaware and trading antiques[395].

Tyrrell and Almeida moved on from coronaviruses because after 229E and OC43 were characterised, the family held little interest for human virologists. Neither of those coronaviruses caused as many severe disease cases as the other known cold viruses, such as influenza, RSV and parainfluenza. While veterinary scientists turned up a steady stream of animal coronaviruses, the list of human coronaviruses remained stuck at two rather uninteresting cold viruses until a new century brought new reasons to pay attention to the family.

The SARS epidemic

On Saturday 22 February 2003, a man vomited on the ninth floor of Hong Kong's Hotel Metropole[396] and kicked coronaviruses to the top of the research agenda.

The man was a doctor based at a hospital in Guangzhou, the capital of the Chinese province of Guangdong, where a mysterious disease had been circulating since November. Not many people had caught the disease but it was bad news for those who did; it infected their lungs to cause a pneumonia that killed around one in every ten of its victims[397]. The people of Guangdong were afraid, and where there is fear, there are rumours. The word was that this mysterious new disease could be cured by white vinegar, sending the price soaring to twelve times its usual value. Another rumour had it that the outbreak stemmed from a bioterrorist attack on Guangzhou's World Trade Centre, which the building management sought to contain by passing vaporised white vinegar through the building's ventilation system[398].

Meanwhile, severe acute respiratory syndrome, or SARS, continued to spread.

Earlier chapters have discussed the role of fine droplets in carrying viruses. Simply breathing is enough to generate a cloud of them. Coughing and sneezing project them much further than exhaling them. But the most efficient way to fill an indoor space with fine droplets is with a really good vomit[399].

That doctor's upchuck infected eight people on the same floor and another two elsewhere in the hotel. The Metropole was a high-end hotel and within the next couple of days, its weekend visitors carried SARS to Singapore, Canada and Vietnam while leaving an outbreak in Hong Kong behind them[400]. Whether or not that traveller from Guangdong recovered is not recorded, but we do know that he inadvertently turned SARS from a provincial outbreak into an international epidemic. In the following months, it spread to 29 countries and killed over 750 people[401].

It also became the target of an international effort to contain the epidemic that was unprecedented in terms of both its speed and the level of coordination between different countries[402]. Public health agencies swung into action to identify anyone showing what might be the first signs of SARS and get them into a hospital isolation room. At the same time, agencies in different countries shared information so they could track down anyone to whom that patient might have given SARS, which sometimes led to infected individuals being traced before they had any idea they were infected.

The public health teams scrambling to contain SARS had a key factor working in their favour: an infected person was only infectious after they fell ill[403], which made it possible to contain the epidemic by making sure anybody who went near a SARS patient took precautions like facemasks and rigorous handwashing. It made SARS a lot easier to contain than pandemics caused by influenza and SARS-CoV-2, which are infectious before they cause any symptoms.

The SARS epidemic ended in July 2003, only five months after it went international in the Hotel Metropole. There are still debates around how much credit to give the international containment effort. Perhaps the containment nipped a pandemic in the bud or perhaps it would have petered out of its own accord. Whichever of those is true, a virus that killed one in every ten people it infected needed to be contained. Even if the epidemic would have ended without those containment measures, limiting the number of people infected certainly saved lives.

The containment effort wasn't the only part of the response to SARS that was unprecedented. The latest refinements in molecular biology techniques enabled virologists to characterise it much faster than any virologist had ever characterised a virus before.

The virologists who got stuck into SARS started with what they called the 'classic'[404] approaches which, by the early twenty-first century, included the electron microscopy techniques that had been so innovative when Almeida introduced them in the 1960s. The first images of the SARS virus capsid revealed the SARS virus to be a member of the coronavirus family[405], closely related to the innocuous cold viruses 229E and OC43.

The next step brought SARS research into the new century. They applied new techniques that showed what made the new virus so viral: its genome. We'll get to where those techniques came from in the next chapter but by May 2003, they were so refined that only three months after that fateful upchuck in Hong Kong, a research group at Vancouver's British Columbia Cancer Agency published the complete genetic sequence of a SARS virus that had ended up in a pair of lungs in Toronto[406].

When that paper came out, a colleague with 20 years of virology behind him gave me a chagrined look and said, 'Not so long ago, that would have kept a dozen virologists in work for years.'

He needn't have worried about virologists being put out to pasture. The process of sequencing a virus's genome might have got much faster but two decades later, there is no shortage of viruses that are worth sequencing.

Meanwhile, the novel techniques that so impressed my colleague were upending the prevailing wisdom regarding coronaviruses. The SARS epidemic showed there was more to the group than a couple of irritating but relatively harmless human cold viruses and a lot of different animal viruses. The questions of where SARS had come from and whether another coronavirus was likely to come from a similar place kicked the coronaviruses to the top of virology's priority list.

Post-SARS coronavirology

In this new era of coronavirus research, one of the first findings was that there were not two cold-causing coronaviruses but four.

In the last two decades of the twentieth century, Rotterdam's Erasmus Medical Centre accumulated a collection of samples containing mysterious viruses. If a patient was hospitalised with what looked like an unidentified virus, it went into a freezer until one of the Erasmus virologists came up with a new plan to identify the virus.

At the dawn of the twenty-first century, Ron Fouchier and his colleagues were having some success using the newly developed array of molecular biology techniques. One of those samples, originally taken from an eight-year-old boy with pneumonia, yielded a virus that infected several different types of cultured cells, including the HeLa cells that were still going strong more than fifty years after Henrietta Lacks's death. So far, so 'classic', but having got the virus replicating in their laboratory, they skipped the electron microscopy and went straight to sequencing the new virus's genome[407]. They found clear similarities with 229E and OC43 but those similarities only went so far. It belonged to the same family, but it was a newly discovered member of that family.

By then, discovering a new virus was a more modest success than it had been in the heady days of the 1960s but it was still noteworthy. They wrote the first draft of a paper but they got distracted when that doctor vomited SARS out of Guangdong and into the world. Fouchier and his colleagues were among many scientists for whom the SARS virus was suddenly the first, last and only priority.

In January 2003, while Fouchier had his hands full with SARS, Lia van der Hoek of the University of Amsterdam managed to culture a virus isolated from a seven-month-old hospitalised with bronchiolitis. She skipped electron microscopy and went straight to an exploration of the virus's genome where she found the same thing that Fouchier had: she was working with a third coronavirus. Moreover, van der Hoek found the same virus in seven other individuals[408], which would lead to it being identified as a cold virus that's as commonplace as 229E and OC43.

Van der Hoek had no idea that she was discovering the same virus described in the manuscript sitting half-forgotten in Fouchier's hard drive less than 100 km away. The SARS epidemic ended and Fouchier became caught up in new priorities, but that paper was not among them.

Fouchier only heard about van der Hoek's coronavirus after she'd submitted her paper[409]. One look at her virus's genome told him she'd discovered the same virus that he'd never got around to telling the world about. Fouchier kicked that draft to the top of his to-do list and got it submitted. As Fouchier put it to me, 'that's how these things go when you find out that you're about to get scooped'[410].

Fouchier was indeed scooped; van der Hoek's paper[411] appeared in March 2004 while Fouchier's[412] didn't appear until the beginning of April. Nevertheless, they are generally regarded as the co-discoverers of the third cold-causing coronavirus, recognising that they both discovered it independently. They agreed to use van der Hoek's name for the new virus: coronavirus NL63. The letters referred to the Netherlands while the number gives a sense of how many viruses a group like hers typically isolates and characterises before finding something new.

Between them, Fouchier and van der Hoek raised the number of cold-causing coronaviruses from two to three, although more detailed analysis of coronavirus genomes revealed that it had not always been so. In the eleventh-century, around the time that King Canute ruled much of the British Isles and Scandinavia, a single coronavirus gave rise to two different ancestors that evolved through centuries of infecting one person after another into the two different coronaviruses NL63 and 229E[413].

In January 2004, while Fouchier and van der Hoek were racing to press, a 71-year-old man was admitted to a Hong Kong hospital with pneumonia. He'd arrived a few days earlier from Shenzhen, a city just across the border between Hong Kong and Guangdong Province. With SARS still on the rampage, it wasn't looking good for a habitual smoker whose lungs still bore the scars of the tuberculosis he'd been treated for some 40 years earlier. Hospital staff took swabs from his nose and his throat and sent them to Patrick Woo at Hong Kong University, presumably expecting them to be laden with SARS virus. However, Woo and his colleagues found that whatever virus might be in those swabs, it wouldn't infect the cell cultures that the SARS virus was happy to replicate in.

Meanwhile, the man was recovering much faster than anyone expected. He was out of hospital within a week which, combined with Woo's inability to culture any virus from those swabs, suggested that whatever put him there wasn't SARS. Yet *something* had caused that pneumonia so Woo set about looking for virus genomes directly from the swabs. He found a member of the coronavirus family and used a novel approach to sequence the virus's genome directly from the tiny amount of genetic material on the swab[414].

June Almeida's immune electron microscopy technique had allowed virologists to get electron microscopy images without having to get the virus into a cell culture. Now Woo and his colleagues had shown it was also possible to elucidate a virus's genome without first culturing the virus. It was a significant step forward from the techniques used to sequence the SARS virus only a year earlier.

That 71-year-old had been hospitalised by a new coronavirus that Woo and his colleagues called HKU1 as an abbreviation for Hong Kong University. Their next question was whether they were looking at a widespread virus that nobody had noticed before or some rarity that had sandbagged a man whose lungs were already half ruined by a lifetime of smoking and tuberculosis. In the aftermath of SARS, Woo had a freezer full of samples from patients who had some sort of respiratory illness. Over 400 of them had shown no sign of the SARS virus and when Woo and his colleagues tested those samples, they found HKU1 in ten of them[415].

We now know that Woo's HKU1 and Fouchier's and van der Hoek's NL63 are both widespread cold viruses that, like all cold viruses, occasionally cause more serious conditions. All three started with samples from a few unfortunate exceptions whose infections had turned serious enough to be mistaken for SARS. As with Robert Chanock's discovery of the parainfluenza viruses and RSV*, it's often the small minority of serious cases that draw a virologist's attention to a cold virus.

A disease that is continuously found in a particular population is called 'endemic' and the four cold-causing coronaviruses have become known as the 'endemic coronaviruses'. Colds are endemic to everyone everywhere, and the quartet of 229E, OC43, NL63 and HKU1 appear

---

* Chapter 8 describes Chanock's seminal work.

to be no more or less widely distributed than any other cold virus. They are four among the more than 200 cold viruses but their ability to infect us repeatedly means that between them, they account for at least one in ten of all the colds that we get[416].

No new endemic coronaviruses have been found since the bonanza year of 2004 but the sudden appearance of SARS opened a whole other question in coronavirus research and now that scientists could quickly sequence a virus's genome, they had the tools to go about asking it: where did all these human coronaviruses come from?

Animal origins

Guangdong Province is home to a thriving trade in wild animals that are trapped and then transported alive to wild animal markets. Many of the first people to catch SARS were involved in the live animal trade[417], suggesting an origin for the SARS virus. With so many coronaviruses infecting so many species of animals, those coronavirus infections put trapped and traded animals firmly – if unintentionally – in the frame as the virus's likely origin.

While Patrick Woo was introducing HKU1 to the world, his Hong Kong University colleague, Leo Poon, was taking the virus hunt across the border to a live animal market in Shenzhen. He found coronaviruses in three palm civets, a raccoon dog and a ferret badger[418].

It proved there were coronaviruses floating around the wild animal trade but for a human virologist like Poon, the important question was whether any of them were finding their way out of the animals and into the people trading those animals. He asked some of the traders for blood samples and found that among 20 animal merchants, eight had antibodies against coronaviruses he'd isolated from those animals. Among 20 vegetable traders who spent less time surrounded by live animals, only one had antibodies against those animals' coronaviruses.

None of the traders had become seriously ill which suggested that the infecting coronaviruses had not been the SARS virus, but Poon had established a clear and worryingly busy coronavirus pathway between animals and humans. He'd also established that SARS was something of an exception; most animal-to-human coronavirus infections were harmless.

However, exceptional events become inevitable if they are given enough opportunities to occur, and a lot of human coronavirus infections amounted to a lot of opportunities for another dangerous exception. If animal-to-human infections were as commonplace across the whole of China's wild animal trade as Poon had found in that one market, there were a worryingly large number of opportunities for another really nasty human coronavirus to infect its way into existence.

Poon's work marked the beginning of a new era in animal coronavirus research. The hunt was on to find out what animals were carrying what coronaviruses and which of them might give rise to the next SARS. Researchers found coronaviruses in mammals of all sorts but one group stood out as being infested with a vast diversity of coronaviruses: the bats[419].

One look at a bat shows us a very peculiar animal that has ended up with both fur and wings with, in many cases, the ability to hear its surroundings in the echoes of its own vocalisations. A bat's peculiarities are more than fur deep. Bats live far longer than other mammals of the same size, they appear to be almost incapable of developing cancers and they carry around a huge diversity of viruses that cause them no ill effects at all[420]. Those viruses include the coronaviruses.

Among the leaders in the quest that led to the bats was Zheng-Li Shi of the Wuhan Institute of Virology, who spent much of the late 2000s and early 2010s hiking around bat roosts in the caves of Southern China. Her colleagues nicknamed her the batwoman[421], but behind the humour lay a serious purpose. She was collecting and sequencing a huge library of bat coronaviruses and it was looking likely that one such coronavirus was the ancestor of SARS.

At some point, a coronavirus had made the jump from a bat to a human. For a coronavirus, that's probably not much of a leap. Just as Poon found evidence that civet and raccoon traders often picked up coronavirus infections from their live commodities, Shi's team found similar infections among people who worked with bats[422] and, also similarly to Poon's findings, none of the infected bat traders had suffered any ill effects from those infections.

It later emerged that SARS probably did not emerge simply through a bat infecting a human. The SARS virus is a mishmash of genetic code from both bat and civet coronaviruses that probably emerged through a particular trick of coronaviruses called recombination: when two different coronaviruses infect the same cell, their genetic code gets mixed up and the viruses that emerge out of that cell are hybrids of the two. It's a process very similar to that which gives rise to pandemic influenza viruses*.

Most of those hybrids will be, in virological terms, a hot mess. Some will be unable to even infect another cell while many that do will not be able to reproduce themselves. The danger is that once in a while, recombination gives rise to a hybrid virus that is not a hot mess at all but one that draws the strengths from both its progenitor viruses and may have properties possessed by neither.

Such was the origin of the SARS virus, which appears to have emerged from a cell in an unfortunate civet that was already carrying a coronavirus of its own when it was infected by a bat virus[423]. That new virus was able to infect a human and, unlike those commonplace coronavirus infections that Poon and Shi found happen all the time, it was able to go on to infect other humans and to cause a disease serious enough to kill one in ten of them.

The more virologists found out about the origin of SARS, the more those virologists worried about the Chinese wild animal trade. The SARS virus came into existence because humans caged bats and civets close enough to infect each other, which simply would not happen in the wild. If it could happen once, it could happen again. Packing different species together allows different species to infect each other with different coronaviruses that may hybridise into some chimaera that could never have existed in nature. Every trader and every buyer at such a market risks being the destination of dozens of complicated pathways from one species to another that could only exist in such a market.

An inconvenient discovery

As techniques for sequencing viruses got faster and researchers around the world expanded the list of known animal coronaviruses, it became apparent that the Chinese wild animal trade was not the only source of human coronaviruses. Comparison of those various sequences revealed

---

* Chapter 6 describes the evolution of pandemic influenza viruses into seasonal influenza viruses.

that OC43 arose as a hybrid between a bat and a cow coronavirus, presumably on some beef or dairy farm. Moreover, OC43 is a relative stripling among the cold viruses. It's so similar to existing cow coronaviruses that the fateful hybridisation probably didn't happen any earlier than the 1870s and may have been substantially later[424].

For 229E, we have less of an idea of when it arose but a much better idea of where: it appears to be a hybrid of bat and camel coronaviruses found in East Africa and the Arabian Peninsula[425]. The route to humans was probably much the same as for OC43: a domestic camel caught a bat coronavirus that hybridised with a coronavirus it was already carrying and infected one of its handlers.

It was a route from animal to human that would be followed by a far more dangerous virus.

On 13 June 2012, a 60-year-old man arrived at the Dr Soliman Fakeeh Hospital in Jeddah, Saudi Arabia. He was running a high fever and coughing up a lot of phlegm. The doctors were worried enough to keep him in the hospital but their attempts to find out what was making him ill left them stumped. None of their diagnostic tests revealed any of the infections that typically caused his symptoms. As the patient's condition deteriorated to a full-blown pneumonia, the pressure was on to find out what was making him so ill. In the hospital laboratory, an Egyptian virologist called Ali Mohammed Zaki was sure he was looking for a virus and, sure enough, something from one of the man's samples was causing cultured cells to fuse into syncytia. That was a good indication that a virus from the patient's sample had infected those cells.

Eleven days after the man was admitted, he died. Now that it was too late to help the patient, Zaki was expected to drop the matter. He couldn't help the man now, and there were plenty of patients who needed his skills.

Zaki didn't drop it. He thought a virus that had killed a man warranted further investigation and, as the virus was replicating in his cell cultures, he had something to investigate. Maybe it was some obscure freak infection that had killed the man through some unlikely combination of genetics, opportunity and misfortune that would never bother anyone else again but, Zaki wondered, what if it wasn't?

He contacted Ron Fouchier, whose research group had a substantial track record in identifying new viruses including coronavirus NL63 eight years earlier. Zaki's message arrived in July and received the reply that most of the Erasmus team were on their summer holidays, but they'd have a look when they got back to work. Meanwhile, Zaki kept at it by himself.

By 2012, even the techniques that had enabled such rapid sequencing of the SARS genome a decade earlier were looking a little dated. Zaki was able to look for 'consensus sequences' within his virus's genome: sections of genetic code that are found in every virus of a given family and no virus of any other family. Once he identified the family, he could then test for more specific sections of genetic code that identified the individual viruses within that family.

Ruling out a viruses a family at a time was less laborious than ruling out each individual virus at a time, but it wasn't quick. This wasn't the coordinated international operation that had identified SARS. This was one man pursuing something he thought might be important.

After drawing a few blanks, Zaki found that the virus that killed the man belonged to the coronavirus family. That was as far as he was equipped to take things by himself, which presented him with some difficult choices. It wasn't likely that one of the four cold-causing coronaviruses had killed an otherwise healthy 60-year-old. The man's symptoms had looked a lot like SARS but there hadn't been a SARS case for years and if there was a new outbreak, there would have been more than one case. The most likely explanation was that Zaki was culturing a hitherto unknown coronavirus.

Without knowing how widely the virus had spread, he had no way to assess how dangerous it was. The man it killed might have been one unfortunate among thousands of infections too mild to have drawn attention. However, the patient he'd cultured the virus from might be the first fatality in an outbreak of a new virus as dangerous as SARS.

In most countries, a scientist making such a discovery could expect some approbation from his peers and possibly a few words of congratulation from a mid-level government functionary. Saudi Arabia was not such a country. Zaki was painfully aware that the Saudi government would not appreciate any suggestion that they might be ground zero for an epidemic. Nevertheless, he sent his samples to Fouchier in Rotterdam and let the chips fall where they may.

Fouchier and his colleagues found that it was indeed a new virus[426] and it proved to be as dangerous as Zaki believed it to be. The disease that's now called Middle Eastern Respiratory Syndrome, or MERS, has caused several outbreaks since then and has killed more than a third of the people known to have been infected with it[427]. That estimate would make it considerably more deadly than SARS but there is a caveat: it is possible that some people it infects don't get ill enough to see a doctor about it and recover without anyone knowing they had it, in which case that one-in-three fatality rate is an exaggeration. However, there can be a fine line between a caveat and a quibble. There's no doubt that MERS is a very dangerous disease.

Zaki's thanks for warning of it was the ire of the Saudi Ministry of Health, which obliged him to get out of the country so fast that he left all his possessions behind him. He returned to Egypt, where he became a professor at Cairo's Ain Shams University[428].

Subsequent research established that the MERS virus has a similar ancestry to 229E: it was a hybrid of a bat and a camel coronavirus that had been circulating in domestic camels for some time but in 2012, something changed that not only allowed it to infect humans from camels but also to allow it to transmit between humans, albeit in a limited enough way that allows outbreaks to be contained[429]. As I write this*, there have been more than 2,500 confirmed cases of MERS and the death toll is climbing toward 1,000[430]. Right now, as you read this, there will inevitably have been more infections and more deaths.

Endemic and emergent coronaviruses

By the mid-2010s, virologists were assigning the coronaviruses to two groups. The four 'endemic coronaviruses' are typical cold viruses, infecting us repeatedly but rarely doing anything worse than making us sniffle and sneeze for a few days. The SARS and MERS viruses were called 'emergent' coronaviruses, which infect humans from animals – once in the case of SARS and repeatedly in the case of MERS – and cause a disease far more serious than a cold.

The differences between the emergent and the endemic coronaviruses do not end with how likely they are to kill you. Another key difference is that like any other cold virus, the endemics become infectious before they cause any illness. Like SARS, MERS only becomes

---

* At the time of writing, the most recent update was posted in May 2024.

infectious after it makes someone ill enough to attract the attention of healthcare professionals trying to contain an outbreak. For SARS and MERS, the relatively late onset of infectiousness proved something of an Achilles heel because once health authorities know one of them is on the rampage, anyone who catches it can be identified and quarantined before they infect anyone else. Such an approach was the basis for stamping out SARS back in 2004 and by the late 2010s, many Middle Eastern public health authorities had more experience than they might have wanted at containing MERS outbreaks.

The dichotomy between the emergent and the endemic coronaviruses raises the question of whether SARS and MERS were the first emergent coronaviruses to cause serious diseases. It's hard to believe that no such viruses existed until the beginning of the twenty-first century and then two burst onto the scene within ten years of each other. The explanation probably lies in virologists' ability to detect them. As we've followed the history of the discovery of the cold viruses, we've seen virology techniques evolve from squirting one person's filtered nasal wash up someone else's nose through to culturing viruses in live cells, visualising them with electron microscopes through to the ability to directly detect a viral genome. It was a battery of approaches that only came of age at the end of the twentieth century, just in time for the 2003 SARS outbreak. History is littered with accounts of epidemics and pandemics of unexplained origin, so it's very likely that some were caused by emergent coronaviruses.

But if nasty coronaviruses have indeed been emerging through the centuries, where did they all go? We can only speculate, but the history of the SARS outbreak gives us a clue because it's tied into our modern world of international travel. In an earlier time, it might never have found its way into a Hong Kong hotel full of international travellers and it might have remained a relatively local outbreak until it ran out of people to infect.

Another possibility is that the emergent coronaviruses of the past might have evolved into the endemic coronaviruses of today. The endemics must have emerged at some point and the in-depth research into OC43 showed that the emergence happened relatively recently. However, there's no firm evidence that they were any nastier when they first emerged than they are now. They might have been the cold viruses we know today since their first hybrid ancestors infected their first humans.

Neither explanation is particularly satisfactory, especially as neither fits the pattern we see with MERS, which constantly circulates in camels and periodically infects a human to cause a new outbreak. It shows no sign of ceasing its continuous emergences or establishing itself as endemic in humans.

Amid all the speculation about coronavirus emergences of the past was a certainty about coronavirus emergences of the future: they were inevitable. As we now know to our cost, there wouldn't be long to wait before the next one.

But before we get to the big one, we should take a deeper look at the molecular techniques that revealed the coronavirus discoveries of Woo, Fouchier, van der Hoek and Zaki. They are now the most important tools in virology and they were crafted by some of the most colourful characters who ever picked up a pipette.

Chapter 10

Virus hunting in the molecular age

David Tyrrell's and June Almeida's discovery of the coronaviruses marked the end of the first boom of cold virus discoveries. For the following three decades, every new cold virus belonged to a familiar family. Virologists were reasonably sure there were more cold virus families to find but they evaded the best efforts of cell culturists and electron microscopists to find them.

Once again, the discovery that broke the deadlock happened to coincide with an arbitrarily significant moment in the calendar: the new millennium.

As the world recovered from the party, Ron Fouchier and his colleagues at Rotterdam's Erasmus Medical Centre got back to working through their collection of samples from people with mysterious respiratory tract infections. One of those colleagues was Bernadette van den Hoogen, who was able to try those samples with a range of cell cultures so broad that it would have brought a tear to George Gey's eye back when he was coaxing Henrietta Lacks's tumour cells into immortality.

Something from one of those mysterious samples started breaking down the cells in some of van den Hoogen's cultures. The cell membranes collapsed and fused, forming the giant masses of protoplasm called syncytia as they died. It looked remarkably similar to what happened to cultured cells infected with the respiratory syncytial virus*, which was a good sign that there was a virus replicating in those cells. Despite the similarities, van den Hoogen was fairly sure it was not in fact RSV. Whatever the new virus was doing, it was doing it far more

---

* Respiratory syncytial virus is one of the cold viruses, described in Chapter 8.

slowly than RSV and it would not replicate at all in cells that readily played host to RSV.

Her next step was to try to get a look at what virus might be in there. Following the techniques pioneered by June Almeida, van den Hoogen and her colleagues imaged their virus under an electron microscope. It did indeed look remarkably like RSV[431].

A few years earlier, that was as far as they could have gone. They would have had to decide whether to call their virus a new type of RSV or a related but essentially different virus, but it would have been little more than an educated guess.

In the new millennium, however, they had a new research tool that enabled them to look beyond the indirect effects of the virus on the cells it infects, beyond the shape of the protein coat that carries it between those cells and into the essence of the virus itself: the sequence of its genome.

The laboratory tools needed to study a virus's genome had been a long time in the making. They originated from a breakthrough made back in 1953, the year in which Maurice Hilleman and Jacqueline Werner discovered the adenovirus. Getting from that breakthrough to the practical technique that van den Hoogen used took nearly fifty years, and illustrates the way that science sometimes progresses despite the personalities involved as much as because of them.

The code of life

Scientific papers vary in their significance. Most add some tiny increment to the sum of human knowledge but will only be read by researchers working on the subject of the paper. A few throw some light on a broader subject area. Now and again, a paper comes along that is so significant that it changes not only the way that scientists view science but how humans view humanity.

On 25 April 1953, issue 4356 of *Nature*, one of the foremost scientific journals, published three such papers across a mere five pages. The first is by far the most widely quoted: a description of the structure of deoxyribonucleic acid, better known by its abbreviation of DNA, by James Watson and Francis Crick of Cambridge University[432]. It was followed by a paper led by Maurice Wilkins of King's College London, whose co-authors included most of the team that generated most of the laboratory data[433]. A conspicuous absence from Wilkins's co-authors was the researcher who developed the techniques that underpinned all three papers. Rosalind Franklin and her postgraduate student, Raymond Gosling, authored the third of the trio[434].

There had been some bitter disputes between the authors, but their papers were scrupulously polite. They generously acknowledged each other's contributions, albeit with one pointed omission: Franklin wrote that she was 'very grateful' to Wilkins and Crick while pointedly omitting Watson. Since then, animosity between Watson and Franklin has become so well-known that it's been acted on screen by Jeff Goldblum and Juliet Stevenson[435] which, if nothing else, indicates how important their description of DNA was.

That legend originated from a single source that is, unfortunately, not particularly reliable: Watson's bestselling memoir, *The Double Helix*[436]. Watson portrayed Franklin as a witch-like figure, lurking in a basement laboratory from which she aggressively repelled all interlopers while she concocted experimental results she couldn't understand.

A key dramatic beat in Watson's account describes him stepping into Franklin's lair for a conversation that ended when she sent him away with a flea in his ear. Watson backed away and bumped into Maurice Wilkins, whom Watson portrayed as downright hag-ridden. They got talking and Wilkins showed Watson one of Franklin's results without her knowledge. One glance was all the peerlessly brilliant Watson needed to perceive details that had eluded both Franklin and Wilkins.

The story of Wilkins giving away Franklin's data without her knowledge has become almost legendary despite Watson's account not making much sense. For one thing, the fierce harridan of Watson's portrayal is at odds with every other description by Franklin's colleagues[437], including Crick, who described her as 'friendly and relaxed'[438].

More importantly, Franklin's 1953 paper shows that she was perfectly capable of interpreting her results. That will surprise nobody who has ever developed a laboratory technique. There is always a process of trial and error that depends on being able to recognise the errors and learn from them. Franklin could never have produced usable results if she hadn't understood them. Moreover, both she and Wilkins were more conversant with the X-ray crystallography techniques she was using than Watson. It's simply not credible that Watson's one glance told him something that had eluded both of them.

The much more prosaic truth was buried in Franklin's notes and papers, which were archived after she died of cancer in 1958, when she was only 37 years old. It was only exhumed in 2023 when historians combed through the archive and found that her results had been shared with Watson and Crick in an informal report. Far from having her work stolen by Watson, Franklin had willingly shared her results with him and Crick[439].

Quite why Watson portrayed himself as the perpetrator of a scientific heist was known only to Watson, but it is striking that the version in *The Double Helix* arrogated the credit for the critical insight to Watson and Watson alone. Since then, many self-appointed defenders of Franklin's legacy have derided Watson and Crick for stealing Franklin's results, apparently oblivious to the fact that they may well be amplifying Watson's self-mythologising. Watson's version and the popular backlash rebound rather unfairly on Crick's legacy; all Crick did was read Franklin's report.

Watson always had his peculiarities. In later life, he took to making public pronouncements about racial determinism which came to a head in 2007, when he told an interviewer he was pessimistic about the future for African people because 'all our social policies are based on the fact that their intelligence is the same as ours – whereas all the testing says not really'[440]. His ill-considered statement got him removed from his position as chancellor of the Watson School of Biological Sciences[441], making him one of the few people to become too unpalatable for a department bearing his own name. It also earned him his place among the six case studies of Nobel laureates in a *Skeptical Inquirer* article describing the 'Nobel disease', in which a minority of laureates use the Nobel Prize as a platform for some very peculiar beliefs. The article argues that the intelligence needed for the good ideas that win Nobel prizes does not always come with the critical thinking skills that protect against becoming absorbed by bad ideas[442].

Lost in the controversy about who contributed what is a key feature of DNA that first appeared in Watson and Crick's 1953 paper, but neither Franklin's nor Wilkins's. It takes the form of two long chains of a small number of subunits entwined in the structure that Watson used for the title of his memoir: the iconic double helix.

When the code of life goes rogue

The double helix of DNA arises out of the very simple structure of the subunit chain that forms the DNA molecule: each link in the chain is one of only four possible subunits. When forming the chain, each of those subunits may link to any of the others but once it is locked into the chain, it may crosslink with only one other in the complementary chain. The sequence of molecules in the chain is like an alphabet with four letters and those four letters write the code that all biology stems from.

The DNA molecule carries the code but to read it, that code is transcribed into a structurally very similar molecule called RNA, which is assembled from its own subunits in a way that complements the gene sequences coded into the DNA molecule.

There is then a second stage of interpretation in which the RNA sequence is translated into a protein molecule. Like DNA and RNA, proteins are made up of subunits arranged into a chain, although the 22 amino acids[443] that are the subunits of a protein molecule are fundamentally different to the subunits of DNA and RNA.

The proteins are the workhorses of biology, taking many different forms and fulfilling many different functions. Some metabolise the nutrients that drive the cell, some form structures that make up the cell, some build other types of molecules that form the structures that shape living things, some are involved in the processes that interpret DNA into RNA into more proteins.

If your childhood memories include your mother admonishing you to eat more protein, she had good reason to say that: the protein we eat is broken down into the amino acids we need to make the proteins of our own. Our cells wouldn't last long without it.

Following those three publications of 1953, DNA became a major focus of biological research. Rosalind Franklin became an early leader in that research after she left King's College to head up her own research group at Birkbeck. She took some of the first steps from being able to describe the molecular structure of DNA to understanding the complex pathway from DNA to RNA to proteins to everything else involved in making life alive. She also took the first steps toward understanding how viruses co-opt the process[444].

A virus is a piece of rogue genetic material that inserts itself into the complex process that turns genes into proteins. It may be a piece of DNA that hijacks the cell's RNA to make proteins for its own ends. It may be a piece of RNA that skips the DNA step altogether and gets straight to assembling proteins out of amino acids intended for the cell's protein production. The specifics of the process vary from one virus to another, but every virus is essentially a DNA or RNA sequence for a genetic code that subverts a cell's metabolism.

Viruses have been tinkering with cellular metabolism for a very long time. Probably for almost as long as there have been cells. Biologists are newcomers to the game who could only get started in 1953, when those three papers described how a cell's metabolism is encoded into DNA. It followed that being able to read that code would reveal the basis for all of life.

The early practitioners of DNA and RNA research must have got a little carried away with their favourite molecule because they called their sub-discipline molecular biology, apparently forgetting that every biological structure is made of molecules of one sort or another.

The first molecular biologists soon found that it's one thing to know the structure of the molecules that carry the code. It's quite another to be able to read it.

Three decades after the structure was first published, the best technique for looking for a given sequence within a DNA molecule was the painfully slow Southern blot. It exploited the fact that any given DNA sequence will bind a complementary sequence. If a complementary 'primer' was synthesised and bound to a trace amount of a radioactive element, the primer would bind to the target sequence in the sample which could be detected by the pattern it formed on an X-ray film. The catch was that in most samples, the target sequence was present in such a tiny quantity that it took three weeks to develop the film.

At that rate, the secrets of life would remain secret for a long time to come.

The PCR epiphany

One Friday evening in the spring of 1983, Kary Mullis was driving home from work when he was struck by a thought. Instead of trying to identify the tiny sliver of a given sequence in a sample, might it be possible to selectively replicate that sequence? With much more of the target sequence in the sample, he wouldn't need to wait three weeks to detect it.

Mullis pulled over and contemplated his insight under a moonlit Californian night while the buckeyes bloomed on the verge and his girlfriend slept in the passenger seat – or possibly, Mullis later wrote, feigned sleep to avoid talking to him.

He eventually drove home with a general idea in his mind. Next Monday, he started work on inventing what would later be called the polymerase chain reaction, usually abbreviated to PCR. It would be a slow and sometimes painful process but within ten years, the idea that came to Mullis by Californian moonlight would revolutionise molecular biology.

At least, that's how Mullis told the story[445]. Mullis was as inclined as James Watson to self-mythologise and his tale of epiphany by moonlight and buckeye blossom sounds suspiciously poetic.

Mullis was the embodiment of the 'Pauling dictum', named for the chemist Linus Pauling's statement that, 'the best way to have a good idea is to have lots of ideas'[446]. Both Pauling* and Mullis would join James Watson in a *Skeptical Inquirer* article describing the 'Nobel disease'[447], which is not a ringing endorsement of the Pauling dictum.

Mullis's first paper epitomised his uncritical excitement over his own ideas. It appeared in *Nature*, the same journal that had published the DNA trifecta, while he was a postgraduate biochemistry student at the University of California in Berkeley. The article had nothing to do with biochemistry but rather hypothesised that half the particles in the universe may be going backward in time while the other half are going forward[448]. At least, that's what I think he hypothesised. It reads like the ramblings of a student who dropped a tab of acid before sitting through a cosmology lecture which, given Mullis's love of LSD[449], may well have been what happened. It's the most bonkers *Nature* paper I've ever read – including the 1901 paper that laid out the basis of eugenics[450], which is at least coherent enough to reveal the flaws in its reasoning – and it left me wondering whether *Nature*'s editorial board were passing

---

* Linus Pauling's demolition of his own legacy with his vitamin C fixation is described in Chapter 12.

around the tabs themselves. In 1968, with the Prague Spring following the previous year's Summer of Love, it wouldn't have been the strangest thing going on in the world.

At Berkeley, Mullis helped a fellow student, Tom White, repair White's broken-down Volkswagen Beetle and made a friend who was far more important to his subsequent career than Mullis ever liked to admit. White helped Mullis graduate by, in White's words, 'cutting all the whacko stuff'[451] out of Mullis's thesis.

After they graduated, White joined the Peace Corps and spent several years training teachers in Liberia while Mullis drifted between jobs and marriages. In 1977, White bumped into Mullis managing his first ex-wife's café in the aftermath of his second divorce[452]. They kept in touch and when White joined a company called Cetus, he got Mullis a job there.

Cetus was one of the first wave of biotechnology start-up companies which were then springing up in and around San Francisco. In the past, pharmaceutical or biotechnology companies had been founded to manufacture established products. If they were successful, they used the profits from that product to fund the research and development that led to future products.

Startups like Cetus began with research and development and placed manufacturing second. Their operating costs were paid not by customers buying products but by investors buying the promise of massive returns when that research and development brought new products to market.

For scientists like Tom White, Cetus promised research opportunities that had previously been the exclusive domain of university departments, but without the bureaucratic hoops academic researchers have to jump through to get their work funded. The lack of bureaucracy extended to a relaxed approach to appointing new staff. When White wanted to give Mullis a job synthesising fragments of DNA called oligonucleotides, Mullis's complete lack of experience working with DNA was less important than that White 'knew he was a good chemist because he'd been synthesising hallucinogenic drugs at Berkeley'[453].

Initially, Mullis didn't disappoint. He developed a reputation for being able to synthesise DNA oligonucleotides faster than anyone else, so when White was promoted to Cetus's head of molecular biology in 1981, he promoted Mullis to head the DNA synthesis laboratory.

It was a decision that would test their friendship to the limit because promotion brought out Mullis's chaotic streak.

Tom White's loyalty

The departments using Mullis's oligonucleotides complained that the speed with which he produced them came at the expense of quality. Mullis responded that any faults lay with everybody else's procedures and not with his nucleotides – and it definitely didn't have anything to do with his staff being distracted by the interpersonal drama that he created.

One evening, a woman who worked in Mullis's laboratory phoned White to say that Mullis had become so jealous of another colleague's supposed interest in her that he was threatening to bring a gun to work. White's response was to cover for Mullis. He placed the man who was being threatened on a mandatory week's leave while Mullis calmed down[454].

Mullis introduced Cetus to PCR in a presentation that didn't go down well. Some of his colleagues were so unimpressed that they walked out. Mullis often came up with what White called 'whacko new ideas'[455] but then, wild ideas are nothing unusual in scientific discussions. Critical thinking is essential to refining ideas but it's no help without ideas to refine. Walking out makes a very strong statement which probably had more to do with the personal animosity that Mullis generated than the whacko-ness of his ideas.

Not everyone dismissed Mullis's idea out of hand. One of his DNA team later said, 'Everyone had this feeling that there is something that they are not quite smart enough to see, but there is a reason why this wouldn't work. Including Kary. Why hasn't someone else thought of this?'[456]

None of the Cetus scientists appear to have known that someone else *had* thought of it. More than ten years earlier, Har Gobind Khorana and his colleagues at the University of Wisconsin-Madison had developed a method for selectively replicating DNA sequences[457] that was very similar to Mullis's 'whacko idea'. If Mullis or anyone else at Cetus had read Khorana's paper, they could have saved themselves a lot of time and effort by using it as a starting point.

Instead, Mullis set about realising his idea from scratch, but his approach to running an experiment was as chaotic as his approach to running a laboratory team.

When a biochemist is developing an experiment, they mix different chemicals in different quantities under different physical conditions and hope they will react with each other to produce the result that they expect. In Mullis's case, the result would look like a dark band on a sheet of a gel-like substance called polyacrylamide. The difficulty was that Mullis couldn't see whether the chemicals were interacting in the way he thought they would. He could only see if the band appeared.

In my own days as a laboratory scientist, I often found myself repeating the truism that nothing ever works first time and if something looks like it worked first time, it will never work again. A scientist's initial idea is only ever a starting point. They don't expect their first attempt to work so they design it to reveal why it doesn't work and point them in the direction of the next thing to try. Realising an idea demands ongoing critical appraisal of one's own work, trying and failing in a way that reveals why the experiment has failed, then changing the chemicals and conditions to fail better until one can finally be confident that the invisible processes happening on the laboratory bench are the same as the processes happening in one's mind.

Critical appraisal was not Mullis's metier. He threw chemicals together, produced something on the gel that was more a smear than a band and announced that he'd made it work. When his colleagues, including White, insisted that a mark on a gel meant nothing unless Mullis could show how it got there, Mullis simply did the same thing again.

Suggestions and searching comments from colleagues are part of the scientific process. I can think of many times when I've become so fixated on the trees of a project that I've lost my bearings in the forest. Like most scientists, I don't exactly enjoy having a colleague point out that I've lost my perspective or that there's a critical flaw in my reasoning but, like most scientists, I can also remember many times when such conversations were exactly what I needed to reorientate myself toward a successful outcome.

Mullis wasn't most scientists. He insisted his smears proved he'd got PCR working and took any suggestion to the contrary as a personal slight. His growing resentment came to a head at a margarita party, held in a Monterey hotel to celebrate the end of a conference that Cetus hosted. Margaritas and Mullis proved to be a bad combination. Mullis, who felt his big idea had been sidelined at the conference, got into a shouting match with a colleague that turned into a shoving match. The two had to be physically separated and Mullis stormed off to his room, where he drunk-dialled White several times to call him a 'jerk'[458]. Eventually, even the long-suffering White's patience ran out. White called the hotel's security, who took Mullis on a long walk along the beach to sober him up. That hotel must have had an unusually obliging security team.

The public tantrum nearly got Mullis fired by Cetus's senior management, most of whom believed his big idea was pie in the sky. White, who seemed to regard being called a jerk as simply part of his friendship with Mullis, demurred. At White's urging, Mullis was replaced as the head of the DNA laboratory and assigned to work full-time on PCR. At the same time, White decided that if there was anything in Mullis's idea, he was going to need help to bring it to fruition – whether he wanted it or not.

Mullis had been working with Fred Faloona, who was a quick learner and a skilled technician but didn't have a degree and didn't contribute any ideas of his own. That suited Mullis, who preferred a sidekick[459] to someone who might disagree with him. However, White realised that leaving the thinking to Mullis wasn't working.

White formed a PCR group of Cetus scientists who could apply the critical thinking skills that Mullis lacked. The key members of the group were Henry Erlich and Randall Saiki. Erlich was working on ways to identify sequences within the human genome that didn't involve waiting three weeks for a Southern blot. Saiki was a much more experienced laboratory technician than Faloona, with the skills to design experiments as well as carry them out.

Erlich and Saiki hauled the PCR project out of the rut that Mullis had dug for it. The speed of their progress is simultaneously an endorsement of Mullis's original idea and an indictment of his approach to realising it. In March 1985, two years after Mullis thought of PCR, Cetus's senior management felt confident enough to file a patent. It's unclear who did what to get to that point but it was Saiki, not Mullis, who took the lead in finding a practical application for PCR.

Saiki focused on a problem that was already occupying a lot of Cetus's laboratory time: diagnosing sickle-cell anaemia, a genetic condition that causes a defect in the red blood cells that carry oxygen around the body. The genetic sequence that causes the condition was known but, as with most situations in which a particular sequence needs to be detected, the quantity of that sequence in any given sample was tiny.

Saiki had been trying to find a stronger signal than the radioactive tracer that took three weeks to make a mark on an X-ray film. Now he'd got PCR to work, he could approach the problem from the opposite end. Instead of trying to tag the DNA coding the sickle-cell gene with a stronger signal, he could replicate that tiny sliver of DNA until it wasn't such a tiny amount any more and there was enough to be quickly detected with the existing tracers.

It worked. Using PCR enabled Saiki to detect the sickle-cell gene within a few hours instead of the three weeks needed for a Southern blot. Saiki's success would be important for getting people with sickle-cell anaemia the treatment they needed quickly but perhaps more importantly, to Cetus's board of directors if not to sickle-cell patients, it showed that PCR was more than a curiosity. If it could be used to identify the sickle-cell gene, it could be used to identify any gene that a scientist wanted to look for.

From the various accounts of what was going on in Cetus, it's unclear where Mullis fitted into the PCR project by then but it's very clear that he wasn't happy with wherever it was. He was even less happy when White refused his demand to be the lead author on every paper involving PCR.

White split the credit, assigning Mullis to write a paper describing the fundamentals of PCR while Erlich and Saiki wrote a separate paper describing their PCR-based diagnostic test for sickle-cell anaemia. Mullis would be a co-author of the latter paper but, to his fury, not the lead author.

Saiki and Erlich got to work and in September 1985, they submitted their paper to *Science*, a journal as prestigious as *Nature*.

Mullis procrastinated. White later said that he 'would bring in beautiful prints of Mandelbrot patterns every morning – using up a whopping amount of computer time instead of writing'[460]. By the time Mullis's paper was ready to submit, Saiki and Erlich's paper[461] had stolen his thunder. Their description of the sickle-cell diagnostic test included a full description of PCR as part of their description of a practical diagnostic test.

The top-tier journals rejected Mullis's paper, which effectively repeated Saiki's description of PCR without a practical application. It ended up in *Methods in Enzymology*[462], a solidly respectable middle-tier journal but not one with the prestige of *Science*.

Meanwhile, Mullis had presented the technique at a symposium hosted by James Watson. The presentation and subsequent report[463] established Mullis as the man behind PCR, which nobody at Cetus ever disputed. Nevertheless, Mullis felt he'd been denied the credit he deserved.

His colleagues would have been justified in feeling that if he'd spent the summer making pretty pictures instead of writing his paper, he had only himself to blame. Mullis didn't see it that way. He blamed them for stealing his idea[464].

Mullis and Cetus parted ways in September 1986, several months before his paper was finally published. There are differing accounts as to how the separation played out, more than one of which originated from Mullis himself.

Cetus, meanwhile, had bigger problems. Investors' money was drying up and its few successful inventions weren't making enough to fill the gap. The senior management fell out and Cetus slowly fell apart[465].

In 1991, Cetus was bought by Chiron, a more successful biotech startup, for $660 million. Chiron promptly sold the intellectual property rights to PCR to Roche, a Swiss conglomerate, for a cool $300 million*. It's an indication of how far Cetus took PCR in the five years after Mullis left.

Mullis was furious. He'd received a $10,000† bonus, a few months before he left, but he spent the rest of his life deriding his former

---

* $660 million in 1991 was worth $1.6 billion in 2025[466], which was equivalent to £1.16 billion or €1.36 billion[467], while $300 million in 1991 was worth $711 million in 2025[466] which was equivalent to £518 million or €604 million[467].

† $10,000 in 1986 was worth around $29,500[468] in 2025, which was equivalent to £21,500 or €25,100[469].

colleagues as 'vultures'[470]. He did not mention that developing intellectual property for Cetus was in his job description, that he'd received the largest bonus ever paid to a Cetus employee or that Erlich and Saiki had received a token $1 each. Nor did he mention that the commercial value of PCR lay in the applications that weren't developed until after he left.

As embarrassing as Mullis became for Cetus and later for Roche, which employed White, Erlich and Saiki to continue working on PCR, they left him unchallenged because they were painfully aware that he held a knife to their throats.

Roche had paid nine figures for PCR because if they owned the intellectual property, anyone who subsequently used PCR in a commercial application would have to pay them royalties. If a legal challenge could persuade a court that PCR had not been an original invention, anybody could use PCR without paying royalties and Roche would be $300 million out of pocket. With that sort of money at stake, someone was bound to challenge Roche's monopoly.

There was one sure-fire way to invalidate that patent: persuade the man Cetus acknowledged as the inventor of PCR to state that it had never been his original idea. Roche's management worried that if they refuted Mullis's public griping, they risked goading him into taking such revenge. During one legal dispute, a Cetus lawyer was dispatched to pacify Mullis by pointing out that if the patent was struck down, so was Mullis's reputation as the inventor of PCR[471].

However strained the relationship between Mullis and his former colleagues, Tom White's loyalty remained unshaken. In 1992, he wrote to the Nobel Committee to nominate Mullis, Saiki and Erlich. On being told that only one of the three slots for the 1993 prize remained open, White replied that it should go to Mullis. He later said it was a 'travesty' that Saiki and Erlich weren't standing beside Mullis when he accepted the prize[472] but in 1993, Saiki and Erlich stayed at home while Mullis flew to Stockholm with his third ex-wife.

Inviting Mullis to a high-profile event had its risks. A journalist later asked his third ex-wife whether he behaved himself in Stockholm.

'I think he tried,' she sighed. 'The police came only once.'[473]

The virus hunter's new tool

While Cetus and later Roche were managing Mullis with flattery, their scientists were finding the uses for PCR that justified that $300 million price tag. Among the first of those uses was the detection of viruses.

Cetus scientists reasoned that if PCR made it easier to detect a given genetic sequence in a sample, and a virus is fundamentally a genetic sequence, then PCR could make it much easier to detect a virus in a sample in the same way as Saiki used it to detect the sickle-cell gene. In 1985, while Saiki and Erlich were writing their paper and Mullis was making Mandelbrot patterns, White initiated a project to invent a way to diagnose a recently discovered virus called HIV.

Infection with HIV lasts for the rest of the infected person's life and before treatment was available, it would considerably shorten that life. It degrades the immune system into acquired immune deficiency syndrome, or AIDS, leaving the infected person vulnerable to microbes that are no threat to a healthy person.

At the time, HIV had been detected almost exclusively in gay men and Haitian immigrants to the USA. Over the next decade, medical scientists would find that HIV was so prevalent around the world that it was classed as a pandemic but in the mid-1980s, it was very much a niche concern. Cetus's marketing division shared the widely held view that HIV was a 'gay plague' and it simply didn't affect enough people to be the basis of a product worth investing in[474].

Most of the senior management agreed, but White believed HIV was important. He assigned senior technician Shirley Kwok to invent a PCR-based diagnostic test for HIV and in doing so, to introduce PCR into virology.

Part of the reason for picking HIV is that it has a trick that is not shared by any of the cold viruses: it inserts its genetic sequence into the chromosomes of the person it infects. The cells it infects circulate in the bloodstream which means that HIV's genetic code is present in a blood sample from an infected person.

Kwok's task was to find it.

It would be difficult and in 1985, there was good reason to suppose it might be dangerous. We now know that Kwok was in no danger because standard laboratory safety procedures are more than adequate to protect anyone working on HIV but in those early days of HIV research, there was no way to assess the risk of a then-untreatable and almost invariably fatal infection. Taking on that project must have taken courage.

Despite the dangers and detractors, Kwok donned her protective equipment and got her diagnostic test working so quickly that she published it[475] while Mullis's paper was stuck in editorial limbo at *Methods of Enzymology*. Kwok had introduced PCR into the virologist's toolkit.

Kwok joined White, Erlich and Saiki at Roche and kept working on PCR. Today's PCR-based tests can not only establish whether an item of genetic code is present in a sample or not but also measure how much of it is in there. Another key development enabled PCR-based systems to detect RNA as well as DNA, which made virologists studying RNA viruses very happy because they could finally detect the most fundamental element of the viruses they were working with.

The one person who was unhappy with these developments was, perhaps inevitably, Kary Mullis. His Nobel Prize had done nothing to ameliorate his resentment, but it gave him a public platform on which to act it out. A journalist who interviewed him in 1994, the year after he received his Nobel Prize, wrote that he 'does not have conversations; he holds forth, spewing opinions like a whale clearing its blowhole'[476].

By the mid-1990s, Mullis's favourite subject about which everyone else was wrong was HIV. In 1994, he delivered a tirade about his belief that HIV did not cause AIDS to the European Society for Clinical Investigation, despite having been invited to speak about the invention of PCR. The mortified chair later described the talk as 'rambling and … inappropriate'[477], supporting his argument with personal attacks on other scientists instead of evidence and illustrated with slides of nude women[478].

It's tempting to speculate that Mullis's HIV denialism was a facet of his resentment toward Cetus. If he couldn't attack PCR itself without undermining his own reputation, perhaps he tried to take some sort of revenge by attacking its foremost application of diagnosing HIV.

His writing and public speaking on HIV got him noticed. In 2000, the South African government adopted denialism as its public health policy on HIV. With more than one in every ten South Africans living with HIV[479], President Thabo Mbeki invited Mullis to join a panel of prominent HIV deniers to endorse his policy of denying his country's HIV crisis[480]. Mbeki not only refused to commit South African resources to treating HIV but also blocked treatment programmes funded by overseas donors. He relented in 2005, when he finally allowed South Africans to be treated for HIV. By then, the policy that Mullis backed is estimated to have cost over 300,000 lives[481].

Mullis's support for Mbeki's disastrous policies was too much for his most loyal friend and supporter. Tom White said, 'I'll never forgive him for that.'[482]

Meanwhile, Mullis had become a vocal contrarian on a range of topics. His autobiography is peppered with denials of well-established scientific facts, including statements that human activities drive climate change and that CFCs damage the ozone layer, although Mullis had even less knowledge of atmospheric science than of virology. Then again, one chapter describes his kidnapping by an alien disguised as a glowing raccoon which left him so traumatised that he had to restore his mental health by shooting up a forest with his AR-15[483]. Mullis's autobiography may not be the place to look for a deep understanding of our world.

Mullis died in 2019, having spent the last quarter-century of his life spraying opinions like bullets from that AR-15. His opinions were underpinned by a myriad of ideas, none of which were good ones. Mullis was a living, breathing, ranting, shooting demonstration of the limits of the Pauling dictum.

His denialism outlasted him. Recordings of some of his HIV diatribes continue to circulate online with any mention of HIV edited out, supposedly arguing that SARS-CoV-2 does not cause COVID-19. Few of the posters appear to realise that Mullis died several months before there was such a thing as SARS-CoV-2.

The creativity that spawns good ideas is the product of knowledge and critical thought. Mullis didn't engage with the existing body of knowledge and he refused to critically appraise his intuitions, rendering himself incapable of creativity. He thrived when Tom White was there to provide that critical thought, first to get him graduated and again to bring PCR to fruition. Left to his own devices, Mullis became a prime example of 'Nobel disease' alongside James Watson and Linus Pauling[484].

The molecular age of cold research

While Mullis acted out his resentment on the public stage, the first molecular biologists were applying PCR to anything and everything involving DNA or RNA. For most people, the most familiar of those uses is not in biological research but in forensic science. When a crime is solved by DNA being found at a crime scene, that solution involves a forensic scientist amplifying selected sequences using PCR.

Its use in virology is less well known, largely because virology appears in TV dramas far less often than forensics, but PCR is central to both research and in treatment of viruses. When Bernadette van den Hoogen discovered her new virus in 2000, she was using techniques built on the three papers describing the DNA molecule, on Mullis's big idea for detecting sequences within that molecule, on Erlich's and Saiki's work that turned the idea into a viable technique, and on Kwok's introduction of that technique into virology.

Van den Hoogen used those techniques to establish her virus as a member of the same paramyxovirus family as RSV, to which it was so closely related that it later followed RSV into the newly defined pneumovirus family[485]. It was even more similar to a virus that had been isolated from birds, called the avian metapneumovirus, so van den Hoogen called her discovery the human metapneumovirus[486]. The name translates, rather loosely, as beyond-the-breath virus or perhaps, beyond-the-breath poison.

Later research revealed that there are two different types of metapneumovirus. Both behave like typical cold viruses: they repeatedly infect each of us and they rarely cause anything worse than a cold, although they occasionally cause bronchitis or pneumonia serious enough to need hospital treatment[487]. They're not the most frequently encountered of the cold viruses, only accounting for around one cold in 50 and they mostly occur in the winter of the mid and high latitudes[488].

Metapneumovirus was the first cold virus to be characterised by a PCR-based technique – but only just. Van den Hoogen's colleague, Ron Fouchier, was already well on the way to getting coronavirus NL63 out of another sample from the same collection. Before long, Lia van der Hoek was doing the same thing in Amsterdam and Patrick Woo was further refining techniques for genome-based discovery to identify coronavirus HKU1.

Metapneumovirus and the two new coronaviruses had been missed by cold virus hunters for five decades and it's no coincidence that one of them, coronavirus NL63, was discovered by two different groups at the same time. Fouchier and van der Hoek had both used techniques that had only just become available. The broader range of cells that could be cultured was important, but it was the PCR-based techniques, recently adapted for discovering and characterising new viruses, that would usher in the next chapter in the hunt for the cold viruses.

Molecular biologists continued to refine and improve the techniques, leading to the discovery of a group of cold viruses that are somewhat weird. By now, you'll have realised that being weird is a common characteristic of viruses; a somewhat weird virus is very weird indeed.

The bocaviruses

Between 2003 and 2004, Stockholm's Karolinska Institutet was, like Erasmus Medical Centre and Hong Kong University, assembling a collection of samples from people hospitalised with respiratory illnesses that had defied diagnosis.

Tobias Allander and his colleagues set to finding the mysterious viruses lurking in those samples and in the first 48 samples they tested, they found two new viruses. They called them ST1 and ST2, abbreviated from Stockholm samples 1 and 2[489]. The genomes of ST1 and ST2 showed that they belonged to a group called the bocaviruses, which included viruses isolated from cattle and dogs but never before from humans. 'Boca' is an abbreviation of 'bovine' and 'canine'; Greco-Latin portmanteaus had given way to acronyms in virology naming fashions.

The Karolinska team looked in a broader pool of 540 samples and found either ST1 or ST2 in 17 of them. Not a huge percentage but enough to confirm that they belong among the ranks of the common cold viruses.

So far, so normal. The weirdness of the bocavirus lies not in what they do but in what they can't do: they can't reproduce themselves independently. Like other viruses, a bocavirus's genome co-opts the biomolecular machinery of a living cell, but unlike most viruses, it doesn't have all the genes it needs to replicate itself.

That raises the question of how bocaviruses manage to exist at all. They must be able to replicate somehow but in the two decades since the discovery of ST1 and ST2, virologists have found two more human bocaviruses but still haven't managed to answer the question of how they manage it.

Having an incomplete genome makes the bocaviruses weird but not unique. An incomplete genome is a defining feature of the parvoviruses, which is the broader family of which the bocavirus group forms a part. Some of the parvoviruses co-opt not only the biomolecular machinery of the cell they infect but also of another virus replicating in that cell, making them parasites of both the host cell and of that other virus. Since virologists have known to look for human bocaviruses, they've often turned up in people who are concurrently infected with another cold virus, but bocaviruses also turn up in people in whom no other cold virus has been found[490]. However, as I write this in 2025, there are almost certainly cold viruses that have yet to be discovered. The human bocaviruses may depend on one of those hitherto undiscovered viruses.

An alternate explanation is that they can only replicate in cells that are in the process of replicating themselves. Some parvoviruses co-opt the machinery of cell division to fill the gaps in their own genomes[491]. A lot of the cells lining the respiratory tract are constantly dividing, which suggests a more likely solution than depending on a compatible virus happening to infect the same cell of the same person at the same time.

At the time of writing, human bocavirus replication is the subject of a debate that looks unlikely to be resolved any time soon. That it merits such a debate shows that human bocaviruses would probably never have been discovered at all if their discovery had depended on getting them to replicate in a cell culture. We know about them because, for the preceding two decades, scientists had nurtured a vague idea from Kary Mullis's chaotic mind into a way to identify an unknown virus on a swab straight out of a patient, as Patrick Woo did when he identified coronavirus HKU1*.

Virus discovery had come a very long way from the days when Christopher Andrewes and Wilson Smith needed a new batch of ferrets every week to sustain their influenza virus.

---

* Chapter 9 describes Woo's discovery of HKU1.

Chapter 11

COVID-19

On the first weekend of September 2023, novelist Amina Akhtar was at a conference in San Diego. Families were enjoying the long Labor Day weekend and her fellow authors were arriving from across the USA, creating an atmosphere that made it easy to forget about COVID-19.

The worst of the pandemic was in the past by then. When COVID-19 was discussed at all, most people assumed that the relatively 'mild' Omicron variant that now dominated the circulating SARS-CoV-2* heralded the end of any serious danger.

Akhtar had accepted every vaccine she'd been offered but she wasn't ready to embrace the general complacency. She was very selective about which panels she attended and wore a respirator mask† as much as possible.

The talk about Omicron being less severe than the earlier variants often missed the point that it's far more infectious. So infectious that the Omicron variant is among the most infectious of all of the infectious diseases[492], and Akhtar couldn't wear her mask all the time. If nothing else, she had to eat and drink.

She started to feel ill on her flight home to Arizona and, as she put it to me, 'that was it. I was in bed and sick right away, and I just did not get better.'[493]

---

* COVID-19 is a somewhat tortured abbreviation of 'coronavirus disease 2019'. The virus that causes it is severe acute respiratory syndrome coronavirus 2, abbreviated to SARS-CoV-2. The '2' distinguishes it from the closely related SARS virus that appeared in 2002, which is SARS-CoV-1 or simply SARS-CoV. Replacing Greco-Latin portmanteaus with acronyms has not made virological nomenclature any more elegant.

† Chapter 13 describes the different types of facemask.

Two years after she caught the Omicron variant of COVID-19 at the age of 45, she remained so debilitated that she has to plan every day around a single activity and then, 'there will be literally not one other thing I'm going to do today … Two days ago, I walked my dog, I got groceries and I made dinner – and I had to pass out because those three things in one day was too much … those are basic things that any, you know, adult should be able to do, and I can't do those things.'[494].

It later emerged that SARS-CoV-2 had really done a number on her. Two years after the infection, a scan revealed that it had left a blood clot in her brain and that the blood pressure in her head was so high that it had distorted her pituitary gland.[495]

A few years earlier, she'd moved to Arizona to take care of her nonagenarian father. COVID-19 has left her so debilitated that now he has to take care of her.

Warnings fulfilled

The virus that would be called SARS-CoV-2 was simultaneously expected and unexpected.

It was expected because virologists like Zheng-Li Shi* had been warning for years that another SARS or MERS was on its way. It could only be a matter of time before another animal coronavirus caused another outbreak. Shi envisaged a third 'emergent' coronavirus that, like the SARS and MERS viruses, caused a dangerous infection of the lungs but did not transmit very fast.

Meanwhile, influenza experts like Jeffery Taubenberger† had been warning that it was only a matter of time before the right combination of bird and human influenza viruses went into a pig and another pandemic influenza came out. They envisaged a virus that could quickly transmit from one person's upper respiratory tract to another's and could spread downward to cause a dangerous infection of the lungs.

Healthcare systems were better prepared for a pandemic influenza than for a hypothetical coronavirus, not least because they were dealing

---

* Chapter 9 discusses SARS, MERS and the work of virologists like Shi on human infections with animal coronaviruses.

† Chapter 6 discusses the work of virologists like Taubenberger on past and future influenza pandemics.

with seasonal influenza every year. There was an influenza vaccine that could be modified to immunise against a newly emerged strain within a few months, there were antiviral medications that could suppress the influenza virus and any doctor, anywhere, could prescribe antibiotics against the secondary bacterial infections that polished off most of the people who died in the 1918 pandemic[496].

Yet SARS-CoV-2 was unexpected because of the way it combined the elements of the two different warnings. It originated exactly as Shi and her colleagues predicted it would: a bat coronavirus infected a human*. It went on to infect many more people at a wild animal market in Wuhan[497], causing an outbreak that expanded into an epidemic and subsequently, a pandemic.

In many ways, SARS-CoV-2 was behaving more like a pandemic influenza than the SARS and MERS viruses to which it is much more closely related. Like a pandemic influenza, it could infect the upper respiratory tract, making it highly infectious, and it had pandemic influenza's dangerous ability to spread down the respiratory tract to the lungs.

In other ways, it was worse than what either Shi or Taubenberger had feared: it was more infectious than a pandemic influenza and it could cause an often-fatal pneumonia without involving bacteria that could be treated with antibiotics. The battery of antivirals and vaccines prepared for influenza was useless against a coronavirus, leaving hospitals trying to suppress their patients' symptoms by pumping oxygen into their lungs as their lymphocytes battled the virus for their lives.

Sometimes the lymphocytes won the battle and the patients recovered. Sometimes they didn't. In its first year, SARS-CoV-2 killed somewhere between one in every 200 and one in every 50 people it infected[498], placing it in the same league of lethality as the 1918 pandemic influenza virus. Although the highest one-in-50 estimate is still lower than the one-in-40 fatality rate estimated for the 1918 pandemic[499], the headline figures only tell part of the story.

---

* Despite the various lurid conspiracy theories, the balance of evidence overwhelmingly supports the 'natural origin' hypothesis in which one of the many bat-to-human coronavirus infections gave rise to a dangerous and highly infective human virus.

Had the 1918 pandemic influenza virus emerged into the world of 2020, it would have been no picnic, but the death toll would have been far lower than it was in 1918. That's partly because many of the fatalities of 1918 were caused by secondary bacterial infections[500] which, by 2020, could have been treated with antibiotics. It's also because in 1918, most hospitals treated influenza patients by putting them to bed and hoping for the best while in 2020, most hospitals were equipped with ventilators, oxygen on tap and a whole range of other paraphernalia to keep a seriously ill patient alive.

Had SARS-CoV-2 emerged in 1918, the death toll doesn't bear thinking about.

Another key difference lay in the way the different pandemics progressed. Whatever mayhem an influenza pandemic may cause, it has a clear endpoint. Each wave blasts through a given region within a few weeks and none of the twentieth-century pandemics lasted for more than a year before the pandemic influenza virus evolved into a seasonal influenza that only rarely causes serious disease.

The evolution of SARS-CoV-2 followed a very different pattern. When it arrived in a region, it tended to stay there, infecting one person after another, until drastic action was taken to block its transmission. That drastic action usually involved facemasks being mandated in public places and 'lockdown', in which measures to drastically reduce human interaction were introduced to limit how many people were infected.

Different countries emphasised different measures of control. Among the most successful were Japan, South Korea and Taiwan[501], all of which got through the worst of the pandemic with fewer infections than other high-income countries[502] and without resorting to nationwide lockdowns. They achieved that through a combination of restricting international travel, universal facemasking and 'contact tracing' that detected people who had been exposed to infection so they could be instructed and supported to avoid infecting anyone else.

Through most of the first year after it emerged, SARS-CoV-2 didn't change much. The virus circulating in October 2020 was much the same as the virus that had first appeared in Wuhan ten months earlier. For immunologists, that stability was encouraging because it suggested an exploitable vulnerability. If SARS-CoV-2 did not change, then immunity to any SARS-CoV-2 was immunity to all SARS-CoV-2. People who had been infected and recovered would have at least some protection against later infections, although it wasn't immediately clear how complete that protection was. As David Tyrrell showed back in 1980[503], immune protection against the 'endemic' coronaviruses is short-lived[504] and it was too early to tell whether immunity to SARS-CoV-2 would last longer than immunity to any of its cold-causing cousins.

By 2020, the state of the art of molecular biology was such that virologists could continually monitor SARS-CoV-2's stability. Two decades after my colleague had grumbled about the SARS virus having been sequenced within three months, molecular biologists could compare the genomes of COVID-19 viruses straight out of patient samples and upload them to an online database that compiled the evolutionary history of the virus for any virologist, and indeed any person with an internet connection, who cared to look.

For nearly the first year of SARS-CoV-2's existence, there wasn't much evolutionary history to see. The SARS-CoV-2 genome hardly changed until November 2020, when it gave up on stability. In southeast England, a variant appeared with a set of mutations that made it substantially more infectious than the 'ancestral' variant it had evolved from[505]. The Alpha variant, as it was later dubbed, was so infectious that it began to replace the ancestral variant, initially in Britain and subsequently across Europe.

Meanwhile, other similarly more infectious variants started popping up in other parts of the world. In the last weeks of 2020, as vaccines against SARS-CoV-2 were beginning to go into arms around the world, the Beta variant emerged in South Africa and the Gamma variant appeared in Brazil[506].

Things went from bad to worse in April 2021 when the Delta variant emerged in India. Delta was more likely to cause the more serious manifestations of COVID-19 than any of the other variants[507] and it was so infectious that it displaced all of them to become the globally dominant variant only a few months after it first emerged[508].

Global domination did not confer stability. The Omicron variant emerged in Botswana in 2021. It was twice as infectious as the highly infectious Delta variant[509] and it spread so fast that it completely replaced all the other variants. By the middle of 2022, Omicron had dramatically changed the nature of COVID-19.

From global pandemic to the worst of all colds

There are a couple of reasons why Omicron gained its reputation as the 'mild' version of COVID-19. For one thing, most people infected by Omicron had some immunity to SARS-CoV-2, either through vaccination, through infection with an earlier version, or both. For another, Omicron is inherently less able to spread beyond the upper respiratory tract to damage the lungs[510].

The best evidence for Omicron being less dangerous than earlier variants arises from an unfortunate set of circumstances that arose in China. For the first two years of the pandemic, the Chinese government had responded to any hint of COVID-19 in a locale with an extremely restrictive lockdown.

The experience of COVID-19 lockdowns differed enormously from one part of China to another. Some regions were barely affected, which is a testament to how effective the Chinese approach was at detecting and stamping out COVID-19 outbreaks. However, the cost of those regions' freedom from both COVID-19 and lockdown was borne by other regions enduring extremely restrictive lockdowns. In August 2022, all 25 million people who lived in the province of Xinjiang were placed under conditions of near house arrest for almost four months[511]. Eventually, anger and frustration with lockdown overwhelmed fear of COVID-19 and, in a country where protests are extremely rare, placard-wielding marchers appeared on streets across China.

The government responded by devolving the responsibility to respond to COVID-19 to regional legislatures[512]. The decision sent a clear message to those legislatures that central government would not support an unpopular lockdown, ending a COVID-19 control policy that had kept most of the Chinese population uninfected.

Few of the most vulnerable people had been vaccinated[513] which, combined with the low infection rate, meant that the abrupt policy change would let Omicron rip through a population in which very few people had any immunity to SARS-CoV-2. The consequences were predictably bad – but not as bad as if the dominant SARS-CoV-2 had been one of the pre-Omicron variants. In 35 days, spanning December 2022 and January 2023, Omicron infected nine out of every ten people in China, but the death rate was between a third[514] to a fifth[515] of what the ancestral variant did in 2020[516]. That still added up to between one and two million[517] which is, by any estimation, a huge number of people.

Since then, Omicron has continued to evolve, and to evolve fast. At the time of writing, it appears to be evolving more than twice as fast as influenza[518] which was previously considered the champion of high-speed evolution among the cold viruses. Every few months, a new Omicron subvariant sweeps around the world which, in any given location, leads to a new wave of infection. Not that there's ever much respite from SARS-CoV-2; wherever in the world you happen to be, there is always a background level of SARS-CoV-2 infection[519]. While other cold viruses tend to follow a predictable annual pattern in which they are most prevalent in any given location at a particular time of year, Omicron has so far remained utterly indifferent to the calendar and is as likely to peak in the middle of a summer heatwave as during the dark days of winter.

Among the scientists who study SARS-CoV-2, there is an ongoing and occasionally heated debate over whether Omicron will eventually fall into a seasonal pattern and why it has so far refused to do so. The first part of the question can only be definitively answered by the time-honoured approach of wait and see; we won't know if it's going to become seasonal unless and until it becomes seasonal.

As for why it has not become seasonal, one possible reason is simply that Omicron is the most infectious of the cold viruses. It's around ten times more infectious than seasonal influenza[520]. Such a high affinity for its host may leave it unaffected by the factors that dictate low seasons for other viruses.

Another factor is that Omicron is the master of evading the immune response, even by the high standards of the coronaviruses among which re-infecting people who are already immune is something of a family trait. Someone who recovers from an Omicron infection will be left with some immunity but not for long. How long varies from one case to another but most people are likely to be vulnerable to reinfection nine months later[521] and for many, protection will last for an even shorter time. That's why some unfortunates endure several bouts of COVID-19 in a single year.

For most people, those COVID-19 infections will feel like any other cold and, as miserable as that feeling may be, will clear up within a few days or at least a few weeks.

As Amina Akhtar found, it's not always that straightforward.

The long haul begins

When Ziyad Al-Aly was 14 years old, he programmed his beloved Commodore 64 to show time on an analogue clock face[522]. It won him a youth coding competition in his home city of Tripoli, Lebanon, making it a solid first step toward becoming the computer scientist he aspired to be.

When he was 16 years old, his father died of cancer after a long illness. Seeing medical practice first-hand had a profound effect on the young Al-Aly. He dropped his ambition to be a computer scientist and decided to study medicine instead.

Al-Aly followed his aspiration through study at the American University of Beirut and then emigrated to the USA to train as a kidney specialist in the Veterans Affairs Health Care System in St Louis. Veterans Affairs was a fortuitous place for a doctor with coding skills. In the USA, military veterans are entitled to healthcare through a nationalised system that operates separately from the private insurance-funded system that provides healthcare for most of the rest of the country. The Veterans Health Administration records all healthcare data on all the veterans it serves in a single enormous database which, for a doctor who knew how to use a for loop, was simultaneously a playground and a goldmine.

Through the 2010s, Al-Aly's work on the Veterans Affairs database established him as one of the USA's leading epidemiologists and by 2019, he was leading a research group at the University of Washington in St Louis. In 2020, the deluge of critically ill patients threatened to overwhelm every hospital in the USA. The situation called for all hands on deck and Al-Aly heeded the call, leaving his databases for an intensive care ward.

In the early weeks of the pandemic, simply keeping patients alive absorbed everything that doctors like Al-Aly had to give. If a patient recovered sufficiently to go home, that was a success to celebrate when so many patients were dying on the ward. There simply wasn't the attention to spare for what happened next which might be why Al-Aly's interest was sparked by neither his databases nor his patients but by an article in the *New York Times*.

The article told the story of Fiona Lowenstein, a journalist who caught COVID-19 in the first wave of March 2020[523]. Being a healthy 26-year-old, Lowenstein was not a particularly high-risk patient but still ended up spending a week in hospital[524]. It wasn't much fun for Lowenstein but in 2020, any COVID-19 patient discharged in a better state than they arrived in was a success. The following weeks revealed that it was a distinctly qualified success. Lowenstein was beset by a constellation of symptoms including serious indigestion, loss of the sense of smell and intense sinus pressure that came and went according to no discernible schedule.

At the time, COVID-19 severity was measured by the most frequently used measure of disease severity: how many people were dying of it. Death is an unambiguously bad outcome and one that's easy to measure. However, death rates only ever capture part of the story of a disease. An infectious agent serious enough to kill a lot of people is likely to be serious enough to cause lasting harm to even more people. Fiona Lowenstein was unfortunate enough to be among the first to be seriously harmed by SARS-CoV-2.

While she was coming to terms with that harm, Lowenstein got online and found she wasn't alone. People were gathering in social media groups to discuss symptoms that hadn't been noticed by anyone who wasn't specifically looking for them, including the epidemiologists studying COVID-19. Lowenstein introduced their struggles to the world, via the *New York Times*, writing of people in their twenties and thirties who were constantly short of breath or plagued by fevers. Their symptoms lasted weeks after they had supposedly recovered from bouts of COVID-19 which, in many cases, had been classed as 'mild' enough that they had not been hospitalised[525].

It's unusual for a medical condition to be described in a newspaper before it's described in a medical journal but then, 2020 was the year where normality took a holiday. That article snagged Al-Aly's attention. When the immediate crisis abated and he returned to his databases, he went looking for evidence more definitive than conversations in chat groups. He would later describe Lowenstein as the 'index case' of a new clinical condition that, when he started looking for it, was still in need of a name.

A few months after Lowenstein's article was published, an Oregon preschool teacher called Amy Watson set up a Facebook group to connect people with lasting COVID-19 symptoms. While she was searching for a name for the group, she happened to notice her favourite baseball cap on her coffee table. She'd been wearing that cap when she took the test that proved she had COVID-19 and, as it looked like the sort of thing a long-distance truck driver would wear, she decided to call the group 'long-haul covid'[526] and her fellow sufferers 'long-haulers'[527].

Watson's terms started as a private joke but they caught on. There have been various attempts to impose technical terms for the condition since then, but even in technical discussions, the term 'long covid' is used far more frequently than any of the long-winded acronyms that have been coined to replace it.

Watson could teach virologists a thing or two about the art of neologism.

How can one virus do all this mayhem?

While Watson was naming long covid, Al-Aly was trying to build the sort of understanding that could never come from newspaper articles or social media discussions. To fully understand how many people were affected by long covid and how it affected them, he needed a comprehensive set of medical information from a large enough group of people to encompass the full range of differing conditions that could follow a COVID-19 infection. Fortunately, he had one: the Department of Veterans Affairs database that Al-Aly and his team had been working on for years.

Al-Aly started by looking at what happened after a relatively mild bout of COVID-19. He identified more than 70,000 people who had tested positive for COVID-19 but had not been hospitalised. 'Mild' in this context can be a relative term; he was drawing information from some people who had very minor symptoms or no symptoms at all and some who felt like merry hell for weeks. The point of sticking to cases that had not needed hospitalisation was to rule out cases serious enough that a prolonged convalescence might be expected.

Al-Aly and his team then selected another four million people who had never tested positive for COVID-19, enabling them to see what conditions were more likely to appear immediately after COVID-19 than in the general run of a veteran's life.

Many of those 70,000 veterans developed long-term conditions after their supposedly mild bout of COVID-19. So far, so expected; that was why Al-Aly and his team were performing their analysis in the first place. However, they were astounded by how many different types of conditions they found. Respiratory problems that left veterans permanently short of breath were no surprise; COVID-19 is primarily a respiratory disease and reports of such problems had dominated discussions of long covid since Lowenstein first wrote about it. However, veterans were reporting everything from cognitive decline to painful indigestion to repeated urinary tract infections and everything in between. Even more perplexingly, some veterans were developing lifelong conditions like heart conditions or diabetes within a month of being diagnosed with COVID-19[528].

'We saw hits in nearly every organ system,' Al-Aly later told an interviewer. 'I do remember initially telling my team here, this cannot be true ... How can one virus do all of this mayhem throughout the body?'[529]

A short-term bout of COVID-19 was leaving many people with all sorts of different long-term conditions – at least, among those who lived long enough for those conditions to count as long-term. For every two deaths among the veterans who had not had COVID-19, there were three among those who had. Those extra deaths were not because they were dying of COVID-19, at least not directly. Their COVID-19 hadn't even been serious enough to need hospitalisation.

Al-Aly and his team did what good scientists do when they get a surprising result: they checked their work. In a complex analysis of an enormous dataset, there were plenty of opportunities for someone to multiply something that should be added or include something that should be excluded. Yet however hard they scrutinised their work, they could find no sign of a mistake.

'We kept doing it and doing it and doing it and kept coming back with the same exact thing,' Al-Aly said. 'This is the data. The virus is doing this to people.'[530]

Whatever nastiness long covid brings, it's very likely to come with a side order of fatigue and muscle pain. Sometimes the fatigue and muscle pain *are* the form that long covid takes. They, more than anything else, are what keep Amina Akhtar from doing more than one or two things every day.

In day-to-day conversation, 'fatigue' usually means feeling tired but in the medical sense, it's far more pernicious because no amount of rest will shift it. With the fatigue comes the cognitive impairment of 'brain fog', eroding short-term memory and concentration span so that trying to absorb any new piece of information feels like bashing a large square peg into a small round hole.

Anyone who has ever been tired will recognise that sensation of being unable to read a book because you can't remember what happened two paragraphs ago or being unable to follow a film because you can't remember who did what five minutes ago. The difference between tiredness and fatigue lies in whether you can restore your abilities with a good night's sleep. Tired people can. Long-haulers can't.

Akhtar described it to me by reference to Aldous Huxley's *Brave New World*, in which humans are engineered for different cognitive capacities, from the alphas who form the intellectual elite down to the drone-like epsilons designed to be content with the most mundane tasks. Akhtar is too modest to liken herself to an alpha so I'll do it for her: the imagination that conjured the murderously disgruntled fashionista in *Fashion Victim* and the mystified stalking target of *Almost Surely Dead* places her among the most intellectually able minds of Huxley's imagination.

Two years after catching COVID-19, she was writing her fourth novel at a rate of a few paragraphs per day, as long as she did nothing else on the same day. She was far from an epsilon, which she proved by coming up with the Huxleyan analogy, though she sounded as though she felt like it.

'I feel like somebody has done that to my brain,' she told me. 'They have actually dumbed down my brain like I feel my brain is significantly dumber than it was before I had covid.'[531]

When Al-Aly's results were published in April 2021, the sheer range of different conditions that comprised long covid may have surprised scientists but it was not particularly controversial. Doctors all over the world were seeing patients with conditions that had started with COVID-19.

The focus of scientific discussion was around how a virus of the upper respiratory tract – at least in most cases, including those that Al-Aly looked at in his initial study – which is usually cleared by the immune system in a couple of weeks[532] causes the long-lasting multi-organ mayhem that Al-Aly had uncovered.

Mechanisms of mayhem

Scientists looked into the questions and they found answers. So many answers that long covid came to look even more nebulous than when it was a subject of Facebook discussions.

Many of those answers involved oxygen. Being a living human being depends on the stuff which we extract from the air in the lungs, transport around the body in the bloodstream and metabolise in every cell from the scalp to the toes. Long covid appears to interfere with each of those three stages.

If SARS-CoV-2 gets into the lungs, it can damage the alveoli that pass oxygen from the air into the bloodstream. Tissue directly damaged by SARS-CoV-2 infection is replaced by scar tissue, which blocks oxygen uptake[533]. No wonder some people are left permanently short of breath.

It's only in a minority of cases where the infection reaches the lungs, especially since Omicron pushed out the earlier variants in mid-2022, but long covid can interfere with the circulation of blood which carries oxygen around the body. Blood needs to flow freely inside the body and to quickly coagulate into a solid outside it, which involves a biochemical balancing act that SARS-CoV-2 disrupts. Long-haulers often have blood that is a little too ready to coagulate, which is how Akhtar ended up with a blood clot in her brain. Sometimes the problem is not a single clot in one place but many 'microclots' of oxygen-carrying red blood cells coagulated together. Microclots are small enough to pass through the major arteries and veins but block the much narrower capillaries, from which oxygen is extracted from the bloodstream and into the cells. When microclots even partially block those capillaries, the cells those capillaries supply do not get enough oxygen[534]. At the same time, long covid may inflame the lining of those capillaries, making them even narrower and prone to being blocked by microclots that would otherwise have passed through them[535].

In other cases, the bloodstream successfully delivers oxygen to cells that cannot make full use of it. Cells metabolise oxygen in structures called mitochondria but in some long-haulers, that metabolism goes awry. Cells switch to the alternative metabolic system that's usually only used when we're doing something so physically strenuous that our lungs can't draw in enough oxygen to keep us going[536]. When a simple walk down the street engages the metabolic symptoms that would usually kick in for a sprint, that walk is likely to make a long-hauler ache as if they've been sprinting. As Akhtar put it, 'I never understood what bone tired was until I got this.'[537]

As if its multi-pronged attack on oxygen transport isn't enough, long covid may do lasting damage in other ways. In some people, long covid interferes with the membrane that filters blood before it gets into the brain, which can lead to physical damage to the brain which is associated with particularly bad brain fog[538].

Long covid sometimes manifests as autoimmunity, in which some of the immune mechanisms that usually protect the body from invading microbes are redirected against the body itself[539] or it may trigger an ongoing inflammatory response.

The latter mechanism looks similar to the misdirected inflammation that causes the symptoms of a cold. Both are misdirected inflammatory responses that leave someone ill and exhausted, but while the inflammation of a cold is only at its worst in the upper respiratory tract and clears up soon after the cold virus is cleared, the inflammation associated with long covid can occur throughout the body and doesn't go away[540].

There's more to the body than human cells. The typical human body has at least as many bacterial cells as human cells[541] and having the right bacteria in the right places is important for good health. It follows that having the wrong bacteria in the wrong places will lead to bad health but unfortunately, research into the body's microbial communities is at a very early stage. Nobody can definitively state what a healthy or an unhealthy microbial community looks like.

Nevertheless, some researchers have shown that in long-haulers, the microbial community in the intestines looks very different to the microbial community of healthy people[542]. The finding led a team at the Chinese University of Hong Kong to try one of the few long covid treatments that has had some sort of effect; they found that six months of capsules containing some of the species that dominate the intestines in healthy people alleviated long-haulers' symptoms although it fell well short of curing them[543].

No wonder Al-Aly and his team needed some time to convince themselves of the extent of the mayhem caused by a virus that's present in the body for such a brief time.

One possible explanation is that in some cases, the virus's sojourn in the body may not be that brief. Several different studies have revealed SARS-CoV-2 proteins or RNA in the intestines of people who recovered from the worst of their symptoms months earlier. Those fragments could be evidence that SARS-CoV-2 sometimes manages to escape the respiratory tract and establish itself in the intestine where it goes on replicating itself and pumping more and more virus that drives those misdirected inflammatory responses[544].

The evidence that at least some long covid cases are caused by SARS-CoV-2 persisting in the gut is strong but not conclusive. Those RNA and protein fragments have been found in the intestines of many long-haulers but they've also been found in the intestines of people who recovered without developing long covid. Conclusive evidence would take the form of replicating SARS-CoV-2 isolated from the intestines of several long-haulers which, at the time of writing, nobody has succeeded in doing. However, one thing the history of cold research tells us is that isolating a virus is difficult; that nobody has isolated replicating SARS-CoV-2 is not evidence that it isn't there to be isolated.

An alternative approach would be to treat long covid with antivirals that stop SARS-CoV-2 from replicating. Those antivirals work by suppressing viral replication, which is very effective at treating the early stages of COVID-19. It follows that if long covid is being driven by ongoing viral replication in the intestines, or indeed anywhere else in the body, then suppressing the replication will suppress the symptoms of long covid. Several trials are ongoing[545] but as I write this, it's too early to know if any of them have been successful.

Everything that's now understood about long covid is derived from research carried out with the ultimate aim of finding a treatment or, more likely with such a multifaceted condition, a range of treatments aimed at the different facets. Every new piece of understanding brings science closer to those treatments. It all started when long-haulers found each other online and realised they weren't the only ones suffering from it or, as Al-Aly put it, 'patients ... inspired us to pursue this'.[546]

The risk of long covid today

By the end of 2022, COVID-19 remained a problem that was a long way from a solution, but it wasn't the crisis it had been in 2020 and

2021. The Omicron variant had replaced the more virulent variants worldwide and the majority of people in the world now had some sort of immunity, either because they had been infected or because they had been vaccinated.

In May 2023, the World Health Organization announced that COVID-19 had become an 'established and ongoing health issue' instead of a 'public health emergency of international concern'[547]. The announcement stated that 'it is time to transition to long-term management of the COVID-19 pandemic', which was widely misreported as a statement that the WHO was declaring an end to the pandemic. It was, after all, before both Amina Akhtar and Lucy, whose story was described in the introduction, suffered their life-changing bouts of COVID-19.

In 2025, COVID-19 was still killing between 65 and 75 people in Britain every week[548] but by then, it was evident that the number of people whom COVID-19 killed was dwarfed by the number of people left severely debilitated. That raises the question of exactly *how* likely a bout of COVID-19 is to leave long covid in its wake, which is a much easier question to ask than to answer.

The difficulties are twofold: firstly, there's the question of what counts as long covid, which can manifest as anything from occasional fatigue to the sort of life-changing debilitation that leaves a previously healthy woman like Amina Akhtar all but housebound. The second is that for an epidemiologist like Al-Aly to work out what proportion of COVID-19 cases lead to long covid, he needs to know how many people have caught COVID-19 in the first place. As Al-Aly's initial analysis showed, long covid often develops after an infection that wasn't serious enough to land someone in hospital so it follows that they may not have consulted a doctor at all. Since the intensive testing and recording of the first couple of years of the pandemic came to an end, there is no record in most databases.

One solution was the basis of the British Office of National Statistics COVID-19 Infection Survey, whose researchers regularly tested around half a million people for COVID-19 infection and then saw how many of them developed long covid. While other studies were only counting the COVID-19 cases that sent people to a doctor, the ONS study was able to count every COVID-19 infection from those that put someone in a hospital to those that the infected person hadn't even noticed.

Their results were sobering. Among people experiencing their first COVID-19 infection, one in 35 were left with long covid symptoms that were at least bad enough to limit their day-to-day activities. The immunity left by that first infection conferred some protection and with a second infection, the risk was only one in 63. That may be lower but when the risk in question was one of life-changing debilitation, it's not one that many of us would be keen to take.

Arriving at those numbers had involved an enormous operation to repeatedly sample over 300,000 people for more than two years. Good epidemiology is expensive and in early 2023, the British government decided it was too expensive and pulled the plug[549]. Today, we're left with the question of whether a COVID-19 infection is still likely to cause long covid as the infections the ONS considered, which happened between November 2021 and October 2022.

As I write this, there are several reasons to suppose that COVID-19 may not be as likely to cause long covid as it did two years ago. For one thing, the survey started when most infections were caused by the highly virulent Delta variant and ended after it was replaced by Omicron. Delta was far more likely to cause long covid than Omicron[550], so we can reasonably expect the risk of long covid to be lower at a time when Omicron is effectively synonymous with COVID-19 – although that was already the case by the time Akhtar caught it; the risk of long covid from Omicron remains substantial.

For another, far more of us now have some immunity to SARS-CoV-2 from past infections, from vaccination, or from both than was the case in 2021 and 2022. Immunity can't prevent SARS-CoV-2 from infecting someone, but it can be the difference between the infection causing a cold and causing a severe illness that needs hospitalisation[551]. It also reduces the chances of ending up with long covid[552]. However, immunity has its limitations: reducing the chances of severe illness or long covid is very different to preventing them from happening at all.

Even if the long covid risk has dropped to a fraction of what the Office of National Statistics survey reported - and since the survey has ended, there's no way to tell whether it has or hasn't - it remains a significant risk when many people get COVID-19 over and over again and the risk is cumulative. Even if the long covid risk diminishes with every infection, every infection still adds to the overall long covid risk.

Imagine it in terms of money. If you don't have long covid yet, your overall risk is set at a certain price and you don't know what that price is. Your first bout of COVID-19 gave you £100 toward that overall price and you didn't get it, so you know the price is more than £100. Your second bout only gave you £50 but now you have £150 toward that price. Still not enough? Along comes your third bout and you get another £25 so now you have £175 – and so on until the day you accumulate enough money to pay the price and, in the case of many long-haulers, to pay it good and hard.

As long covid remains untreatable, the best way to avoid it is not to get COVID-19. The next section will discuss possible ways to protect ourselves from cold viruses like COVID-19, both as individuals and as a society. Unlike most cold viruses, there are now vaccines against COVID-19 which, while unable to induce enough immunity to completely protect against it, roughly halve the risk of COVID-19 infection leading to long covid[553].

All of which is of limited help to long-haulers because long covid can be a very long haul indeed. It lasts for more than a year in at least nine out of ten long-haulers[554]. Around half recover over two years and two-thirds over three years[555], which means that a third of long-haulers are still hauling after three years. The implication is that the longer long covid has lasted, the longer it is likely to last.

In 2024, Ziyad Al-Aly led a review of the various surveys that estimated that around one in every 15 American adults and one in every hundred American children had long covid[556]. The most recent figure for Britain, published in 2023, did not distinguish between adults and children which is probably why it fell squarely between the American estimates at one person in 35[557].

The situation is not helped by most people, from lay people to public health policymakers, underestimating the danger that SARS-CoV-2 still poses. In most countries, vaccines and antiviral treatments are only available to a minority of people classed as high-risk despite the evidence that both vaccines and antivirals considerably reduce the chances of being left with debilitating long covid[558].

The policy of denial and complacency is a source of immense frustration to medical scientists working on SARS-CoV-2. Raina MacIntyre, an epidemiologist at the University of New South Wales, described the situation to me as a 'mass disabling event'[559] that shows no sign of ending. I put that to Stephen Griffin, a virologist at the University of Leeds and co-chair of the Independent SAGE group that kept Britain informed through the darkest days of the pandemic, who said, 'I think that's fair.'[560]

Amina Akhtar and Lucy are in good company, which is unlikely to comfort either of them or their fellow long-haulers. 'I talk to people,' Akhtar told me[561]. 'Everyone's miserable because nobody knows how to find something to enjoy again ... it's like, everything's been taken away ... How do you start over and find something new, something different and short of just, you know, watching Netflix in bed?'

With SARS-CoV-2 allowed to spread unchecked and long covid treatments remaining elusive, more and more people will be asking that question in the coming years.

That probably makes you wonder how you can avoid being one of them – assuming, of course, that you don't have long covid already. The next section will suggest some answers to that question.

Section 3

What can we do about colds?

Chapter 12

Over-the-counter medications

In November 1924, while Alphonse Dochez was finding out that colds are caused by viruses*, US President Calvin Coolidge got so annoyed with a cold that he had himself shut in a gas chamber. He spent three-quarters of his time studying the perennial reading of American presidents, a draft immigration bill, while army surgeons pumped chlorine into the chamber. The gas was at a lower concentration than had been used to poison soldiers on the Western Front a few years earlier, but those surgeons were nevertheless gassing their commander in chief[562].

Coolidge declared himself cured though if he was, it's unlikely that the chlorine had anything to do with it. Later experiments showed that chlorine did not affect colds.

However, a president's opinion often carries more weight than evidence. A San Francisco firm called Richard, Price and Hyde developed the Chlorine Kilacold Bomb which they promised would completely cure 97.3 per cent of all colds. One can see the advertising department's handiwork in that decimal point. It claims a level of precision that could only be reached by the sort of meticulous experimentation that had, in fact, shown that the figure was 0 per cent.

It goes without saying – hopefully – that it's a bad idea to breathe poison gas to cure colds or for any other reason. Nevertheless, Coolidge's enthusiasm for being gassed illustrates the reasons why every pharmacy has shelves stacked with a bewildering array of cold cures, each of which has its passionate advocates and their scornful naysayers in equal measure.

---

* Chapter 1 describes how Dochez did it.

Perhaps the most important of those reasons is that it's very difficult to measure how bad someone's cold is. A researcher can measure the temperature of someone with a cold, but a cold can make someone feel like merry hell without causing the slightest hint of fever. Another researcher can swab someone's nose and see if there's a virus replicating up there, but a nose can be packed full of happily replicating viruses that don't cause a single symptom. The closest thing to an objective measure of a cold is weighing the mucus produced: supply a cold sufferer with a box of tissues for them to blow their nose with, then weigh the used tissues, subtract the weight of the tissues and you have the weight of the mucus that has been sneezed or blown.

Mucus can be weighed with a reasonable degree of precision but that doesn't make it a particularly good metric. The volume of gunk that comes out of a cold-stricken nose is a poor guide to how miserable a cold makes you feel.

Unless a cold progresses to serious complications like pneumonia or bronchiolitis, it's as nasty as it feels it is. It follows that any cold remedy that makes a cold feel less nasty is, by definition, an effective cold remedy. If Coolidge came out of that gas chamber feeling better, the treatment had been effective.

That's the basis for the placebo effect. Taking positive action induces the secretion of opioid hormones which suppress pain and improve mood[563]. For the cold-ridden Coolidge, that meant that anything he did about the cold made him feel he'd done something about it – even if the something was as unwise as having himself gassed.

As well as giving himself a dose of opioids, Coolidge may also have been experiencing a phenomenon called 'regression to the mean'. His 'mean' was his usual state of well-being which would have been much better than when he had a cold. However, most people start to recover from most colds after a few days, even if they do nothing at all to treat them. If Coolidge lost patience with his cold and had himself gassed at the point at which he was about to start feeling better, he may have attributed his recovery to the chlorine even though his cold would have cleared up just as quickly if he'd read that immigration bill at home while sipping a hot chocolate.

Testing cold remedies

The placebo effect and regression to the mean can make us feel that a treatment is working even if it has no effect on the inflammation or on the virus causing it.

This was the problem facing a team of researchers working for the British Medical Research Council that included Britain's leading cold researcher of the mid-twentieth century, Christopher Andrewes. In the winter cold season of 1943–1944, the MRC team set out to see whether a fungal extract called patulin worked as a cold remedy. The most important results would come from asking their volunteers how they felt which, they realised, would be a poor way to evaluate the treatment if the patulin-treated volunteers believed the patulin was going to help them.

They weren't the first to recognise that the subjects of a clinical trial could be subject to the placebo effect, but those MRC researchers went a step further than any researcher had before: they realised that they themselves would be biased. They had no financial interest in patulin but they were about to dedicate months of their lives to testing it and, like any medical researcher dedicating that sort of time and effort, they would rather spend it proving a product that would help countless people in the future than proving it was useless and belonged in the dustbin.

The patulin researchers worried that their bias would affect how they asked a cold sufferer how they were feeling. If they sounded more hopeful when they asked someone they'd treated than when they asked someone they hadn't, then their tone of voice might elicit a more upbeat answer from the people who had received patulin.

Some of the recruits would be treated with patulin dissolved in saline and squirted up their noses while others would receive a pure saline placebo. The only way either a recruit or a researcher could tell which was which was by the label. The researchers realised that neither the recruits nor any researchers the recruits spoke to needed to know which was which. If the treatment and placebo were labelled with codes known only to researchers who stayed away from the recruits, the effect of patulin would be assessed beneath a veil of unbiased ignorance.

That first 'double-blind placebo-controlled trial' was the most rigorously conducted clinical trial in history and it demonstrated that patulin did not have the slightest effect on the common cold[564]. More importantly, it set a new standard in the rigour with which clinical trials are still conducted today.

For a scientist planning to evaluate a medical intervention, the double-blind placebo-controlled trial is the first choice of study design. Other approaches are only used for situations where practicalities make it impossible. For example, this chapter draws on many double-blind placebo-controlled trials of pills and nasal sprays but there is no placebo control for inhaling steam.

So much for chlorine and patulin. Neither does any good. The rest of this chapter is a whistle-stop tour of the remedies for which we have at least some evidence that they do work.

Before diving in, a few words of explanation are in order. This chapter covers remedies that can be put together out of a kitchen cupboard or bought over the counter at a pharmacy. Consequently, it will not discuss antiviral drugs, each of which targets a narrow group of cold viruses and, in most countries, requires a prescription.

When we buy a cold remedy from a pharmacist, we rarely know which virus is causing the cold. We need something that works against all cold viruses which, in practice, usually means something that suppresses the inflammatory response rather than attacking the virus itself. Throughout this chapter, I'll be calling such pharmaceuticals by their chemical names which you'll find in the list of ingredients on the back of the packet, not the brand name screeching for your attention on the front. Most of those pharmaceuticals are marketed under different brand names at different prices and I have no idea what brands you'll find in your local pharmacy.

There are a few things you won't find here, including most of the cold remedies that don't have reasonably good evidence behind them. If I tried to describe every single cold remedy that has ever been suggested, I'd fill this whole book describing remedies for which there is no evidence of effectiveness. Some have never been evaluated against colds, leaving nothing much to say. Others have been tested in several different studies and found to have no discernible effect on colds. Ginseng[565] and echinacea[566], two of the most widely touted herbal cold remedies, fall squarely into the latter category.

One thing you won't find in this chapter is a miracle cold cure. The best cold remedies will make a cold more bearable and sometimes a little shorter. The only thing that will make a cold go away is the only thing that has ever made a cold go away: the lymphocytes and antibodies of our own adaptive immune response, which take their sweet time about it.

Vitamin C

In 1966, Linus Pauling stated that he wanted to live for another 15 to 20 years. It was a common enough wish for a 65-year-old man but there was nothing commonplace about Pauling. His discipline-blending scientific career led him to a series of breakthroughs in the physical chemistry of complex biomolecules, earning him the Presidential Medal for Merit and the 1954 Nobel Prize in Chemistry Prize. His political campaigning was instrumental to the international agreement to end nuclear weapon testing, earning him the 1962 Nobel Peace Prize and a place on an FBI blacklist[567]. By 1966, Pauling was among the most famous scientists in the USA, which is probably why his offhand comment about longevity received a reply.

Biochemist Irwin Stone wrote to promise Pauling another 50 years of life. All Pauling had to do was swallow a gram or two of vitamin C every day[568]. Stone's letter set Pauling on a path that so tarnished his legacy that he ended up as a case study in 'remarkable lapses of critical thinking'[569] among Nobel laureates, alongside James Watson and the HIV-denying Kary Mullis*.

Stone was plugging what he called 'orthomolecular medicine', which he claimed was a way to stay healthy by varying the concentrations of the various substances present in the human body. It was the latest iteration of an oft-repeated fallacy: that the human body is a mixing bowl that needs only the perfect set of ingredients to function perfectly. Such a view is a grotesque oversimplification. The human body is not a passive receptacle for whatever substances we put into it. It's an incredibly complex and dynamic system that regulates different biochemistries in different parts of the body. Pauling understood that too well to cast aside his half-century-long habit of critical thinking.

---

* Watson's and Mullis's idiosyncrasies are described in Chapter 10.

He started by delving into the technical literature. He found no evidence for Stone's promise that vitamin C would keep anyone alive until the age of 115 but then, he wouldn't have expected to. Nobody had done a 50-year trial of vitamin C supplementation. He did, however, find several assessments of the effect of vitamin C supplementation on colds, which was a place to start. If vitamin C was the wonder substance that Stone claimed it to be, Pauling reasoned that it should have some effect on relatively minor ailments like colds.

Then he ran into a very common problem in evaluating cold remedies: different studies returned different results. Some showed that people taking vitamin C supplements had fewer colds – although the differences they reported were marginal at best – while other studies found that vitamin C had no effect.

In the past, it would have been left to whoever read those studies to draw the conclusion that seemed reasonable which, all too often, ended up being the conclusion that fit the reader's preconceptions. Pauling wanted a more systematic way of going about it.

Four of the studies had used a very similar design, so Pauling combined the results into a single analysis. By treating them as one big study, he had far more data than the investigators of any one of those studies. More data made Pauling's analysis much more powerful in the statistical sense of the word; he would be able to detect an effect too small to have shown up in any one study.

Pauling did not invent 'meta-analysis', as such multi-study analyses are called, but he was the first to apply it to the science of the common cold and he was instrumental in introducing it to medical research more broadly. Meta-analysis is now widely used to compare results from different studies. It's also the cornerstone of the rest of this chapter, much of which discusses cold remedies that have been evaluated in many different trials.

Pauling's meta-analysis led him to conclude that not only did a gram of vitamin C per day halve the number of colds that someone caught, but it also made the colds they did catch substantially less severe[570].

Unfortunately, Pauling drew that conclusion by botching his meta-analysis. He ran his meta-analysis on the studies that gave the most encouraging results while excluding the rest which, he said, recruited too few volunteers to give a definitive result. That may have been a valid critique of each of those studies considered independently but it's no reason to exclude them from a meta-analysis, for which the whole point is to combine many small studies into one big analysis.

For all its flaws, Pauling's conclusion was a gift to pharmaceutical branding. It launched many an advertising campaign promoting vitamin C as an infallible cold cure because, after all, how could a Nobel laureate be wrong – just don't think too hard about Kary Mullis.

Even if Pauling had convinced himself that vitamin C prevents a few colds, he was still a long way from proving it would keep him alive until he was 115. His next move was to try to prove vitamin C could be used to treat cancer but instead of running any sort of clinical trial, he and some colleagues simply gave vitamin C to cancer patients[571] and announced it had prolonged their lives[572].

Pauling's dabbling in cancer research delivered a thorough monstering to his hard-won reputation without proving anything about vitamin C. Nevertheless, his faith in vitamin C remained unshaken until his death at the age of 93. He got more than the 20 years he'd hoped for, although not the 50 years that Stone promised him.

Pauling's advocacy launched the idea of vitamin C as a cold remedy. More recent research has been far more rigorous, not least in the application of the meta-analysis technique that Pauling pioneered. At the time of writing, the most recent meta-analysis – which includes a thorough explanation of how the investigators chose which studies to include – was published in 2013[573]. Drawing on far more studies than were available to Pauling, it showed that vitamin C supplementation did not reduce the number of colds that anyone suffers from.

There is some better news. Taking at least 0.2g* every day, cold or no cold, reduced the duration of colds, albeit not by much. It shortened colds by between 7 and 21 per cent in children and between 3.5 and 12 per cent in adults. In other words, a typical person will recover from a cold a day or two earlier if they take vitamin C supplements.

Moreover, there is a pattern to which studies show an effect of vitamin C and which do not. Some of the strongest effects of vitamin C supplementation were found among British men in the 1970s[574], which

---

* 0.2 grams is 200 milligrams, which is a fifth of the gram per day that Stone persuaded Pauling he needed.

wasn't a demographic known for a healthy diet containing plenty of fresh fruit and vegetables.

The finding that supplements benefited men more than women is not peculiar to 1970s Britain. Men tend to need more vitamin C than women, although pregnant women need more than men[575].

The dietary question was picked up by a research group at New Zealand's University of Otago who set up a trial with an alternative approach to supplementation. Instead of giving their male volunteers vitamin C pills, they dosed them with different quantities of kiwifruit. One wonders how many kiwi jokes they had to endure when presenting their results at international conferences but their results were clear: half a kiwifruit every day was all that was needed to load their muscle and immune cells with all the vitamin C they needed[576]. Any more vitamin C than that was urinated straight out of their bodies.

The Otago researchers chose kiwifruit not out of any sense of botanical patriotism but because of all the widely available fruits, kiwi has one of the highest concentrations of vitamin C. One would need to chomp through much more of any other fruit or vegetable to get the same amount of vitamin C. However, the volunteers were placed on a diet designed to be vitamin C-deficient ensuring that the kiwifruit was their main source. Anyone with a reasonably healthy diet is unlikely to need even half a kiwifruit to top them up.

Pulling all that together, it's evident that someone who isn't getting enough vitamin C is likely to take longer to fight off a cold than someone who is getting plenty. However, as little as half a kiwifruit per day provides plenty of vitamin C without needing supplements from the pharmacy and someone who includes a modicum of fruit and vegetables in their diet is unlikely to need even that.

Vitamin D

Vitamin D is a gift of the sun. When its ultraviolet rays hit human skin, they convert a biochemical with the unlovely name of 7-dehydrocholesterol, which the human body can synthesise unaided, into vitamin D, which it can't[577]. That's not a reason to risk sunburn; a few minutes of direct sunlight every day is enough to synthesise all the vitamin D we need.

If you live in a mid or high latitude, you'll have seen the catch. During the winter, you won't be getting very much sunlight. That's because the atmosphere filters out ultraviolet and the lower the sun is in the sky, the more atmosphere those rays have to push through before they touch your skin. Even on a clear winter's day, next to no ultraviolet light makes it as far as a ground-bound human.

In Britain, a typical mid-latitude country, the winter brings vitamin D deficiency to nearly one in every five people, although the deficiencies are not equally distributed. In Scotland, where the sun spends December looking like it can hardly be bothered to clear the horizon, it's closer to one in every three[578]. Moreover, vitamin D deficiency is somewhat racist in its distribution. The melanin pigment which darkens skin and protects it from ultraviolet also blocks the conversion to vitamin D, making deficiencies around twice as prevalent among black British or people of South Asian origin than among white Britons.

That's bad news for all sorts of reasons. In children, sustained vitamin D deficiency causes the abnormal bone development of rickets. More pertinent to the subject of colds, vitamin D is involved in many of the processes of the immune system.

If we can't get enough sunlight, there's another way to get vitamin D: we can eat it. There's a fair amount of it in some fish and in eggs, which is why not everyone in Britain is vitamin D deficient over the winter. If that doesn't do the trick, any pharmacy will offer vitamin D pills.

As with vitamin C, different trials of vitamin D supplementation have given different results, so we need a meta-analysis to make sense of the available data. Such a meta-analysis was published in 2017, when the authors found that people who took vitamin D supplements did have fewer colds than people who did not – but only just[579]. The difference was so small that it needed a meta-analysis to see it.

When the meta-analysts dug a little deeper and focused on the studies that measured vitamin D levels in the bloodstream, the contradictory results started to make more sense. They found that supplementation only made a difference to people who entered a study deficient in vitamin D. Anyone who already had plenty of vitamin D in their bloodstream gained no benefit from supplements.

The meta-analysts also found that the dose regime matters. Among those people with vitamin D deficiency, daily or weekly doses led to fewer colds but a single large dose, such as the injections that are offered to older people in some countries, did nothing against colds. That's no reason to refuse a vitamin D injection if a doctor recommends it; vitamin D has many different functions and protecting against colds is only one of them.

Zinc

The vitamins are classed as micronutrients, meaning that the human body needs a small amount, and only a small amount, to function properly. Not every micronutrient is a biomolecule as complex as a vitamin. Some are trace elements and the trace element that has received the most attention from cold researchers is zinc.

All cells need a certain amount of zinc and when the lymphocytes mobilise against a cold virus, they need more of it[580]. Several researchers have looked into whether taking zinc in the early stages of a cold can help the lymphocytes clear the virus a little faster.

As with the studies of vitamins, different researchers found different results that needed meta-analysis to untangle. It turns out that zinc can take between one and three days off the duration of a cold[581] – but only if it's taken in a way that tastes awful and leaves a lot of people feeling downright nauseous.

Swallowing zinc in a pill or a syrup doesn't help. The zinc needs to be delivered directly to the infected throat at the first sign of a cold, and at least 75 milligrams per day is needed to have any effect. That's several pills worth of most formulations which, given that a zinc lozenge tastes like a lump of chalk, isn't the most attractive prospect. Several pharmaceutical companies market zinc with flavouring agents or in more palatable forms like zinc gluconate or zinc acetate – all of which negate any benefit conferred by the zinc[582].

When children regularly take zinc lozenges, they have fewer colds and fewer days off school although the effect is small, averaging out to less than a day per school year[583]. None of the published papers explain how to persuade a healthy child to dissolve a nasty-tasting zinc lozenge in their mouths every day.

Nevertheless, there is good evidence that if we start taking 75 milligrams of zinc per day at the first sign of a cold, we'll probably recover slightly faster than if we didn't – if we can put up with the awful taste.

Probiotics

A typical man's body contains as many bacterial cells as human cells while a typical woman's contains twice as many bacterial as human cells[584]. The difference arises because although bacteria coat every surface from the skin to the respiratory tract to the entire digestive tract, it's the lower intestine that holds most of the bacterial cells we carry around with us and women have proportionally larger lower intestines than men.

That doesn't make any of us only half or one-third human. In a typical 70-kilogram man, for example, all those bacteria would add up to a mere fifth of a kilogram. Despite its paltry weight, recent decades have seen an increasing understanding of the ways that the microbiome, to use the technical term for the human microbial community, is essential for health. Unfortunately, what science has learned about the microbiome is still a tiny fraction of what there is to know.

We do know that the human microbiome is enormously diverse. Different species of bacteria make their homes in different parts of the body, but look in the same part of the body of different people and you'll find very different bacteria there.

One study helmed by Ronald Turner, a colleague of Jack Gwaltney's at the University of Virginia in Charlottesville, recruited volunteers to see whether the bacteria that lived up their noses affected what happened when they were deliberately infected with a rhinovirus. Five days after the infection, there were more rhinoviruses up noses that were dominated by a *Moraxella* bacteria but those people didn't have the nastiest colds. Those happened to people whose noses were not dominated by any bacterium in particular[585].

That doesn't mean we should all be stuffing *Moraxella* bacteria up our noses to tone down our colds. Even if ameliorating colds was as simple as cultivating *Moraxella* up our noses – and the microbiome is anything but simple – it's not a good idea because some species of *Moraxella* have a dark side. Some species can cause infections like bronchitis and ear infections.

However, there may be some benefit in trying to adjust our bacterial communities, as shown by the trial that showed that taking bacteria in the form of probiotics helped people with long covid*. Such probiotics are now available from most pharmacists but they usually contain *Bifidobacterium* and *Lactobacillus*, which colonise the intestine rather than the nose.

Nevertheless, *Bifidobacterium* and *Lactobacillus* probiotics do confer some protection against colds, although not very strong protection. A meta-analysis found that people taking probiotics were likely to have slightly fewer colds and that the colds they had were sometimes a day or two shorter[586]. It's a discernible difference but not a very impressive one; it's unlikely that anyone who hadn't volunteered for a research study would have noticed a difference.

The result raises the question of why putting bacteria into the intestine affects an illness of the upper respiratory tract. The most likely explanation is that what's going on in the digestive system affects how likely the rest of the body is to respond to an infection with the sort of inflammation that causes a cold. As science uncovers more of the secrets of the microbiome, we may learn whether that's true or whether there is something else altogether going on.

Painkillers

So far, we've looked at ways to improve the immune response to a cold virus, some of which help but none of which are particularly impressive. Now let's look at an alternative approach. A cold is miserable because of the sore throat, the headache and the general misery. In short, we don't like colds because they hurt.

Can we make it hurt less?

Pharmacies sell a broad range of painkillers intended to do exactly that, some of which are more powerful than others. For colds, we'd be wise to stick to the less powerful ones. We don't want to be using opiates and opioids that mimic the body's opioid hormones because they tend to be addictive. Colds don't hurt enough to risk getting hooked.

Most pharmacies offer three non-addictive painkillers: aspirin, paracetamol† and ibuprofen. All three work by inhibiting the synthesis

---

* The Chinese University of Hong Kong study that showed probiotics alleviate long covid is described in Chapter 12.

† Also known as acetaminophen and usually branded as Tylenol in the USA.

of a hormone called prostaglandin, which causes inflammation and pain[587].

Aspirin and ibuprofen operate throughout the body, suppressing both inflammation and the pain caused by that inflammation. Paracetamol goes straight for the nervous system so it doesn't prevent the inflammation, but it does stop it from hurting.

That's an important difference when we're looking for a cold remedy because it means ibuprofen and aspirin suppress the bradykinin that swells the nasal blood vessels to block the nose while paracetamol, which blocks the sensation of pain rather than the inflammation causing it, does not.

Of the three, aspirin is the last choice for colds. It's less effective than either paracetamol or ibuprofen and in children, it can cause a rare but life-threatening condition called Reye's syndrome[588].

Direct comparisons between paracetamol and ibuprofen have shown they're both fairly effective with little to choose between the two of them[589] – as long as they're used according to the dose regime described in the packet. Causing pain is only one of prostaglandin's functions. Too much of either paracetamol or ibuprofen risks shutting down prostaglandin functions that we can't do without. Overdosing is a very bad idea but as long as we keep to the recommended dose for the few days it takes to recover, both paracetamol and ibuprofen should make the cold more bearable and won't do us any harm.

Decongestants

The bunged-up nose is the most emblematic of all the symptoms of a cold, if only because it's the most visible. It's not obvious when a colleague is struggling to focus while feeling that their throat is being sandpapered and there's a pneumatic drill banging away in their head, but when their voice is distorted into a ventriloquist's parody and they're filling tissues with nasal gunk, it's obvious to anyone in the room. I'm sure I'm not the only one who feels embarrassed when it's my voice that's distorted and my nose that's running like a tap. No wonder pharmacy shelves are groaning with decongestants promising to restore both voice and airflow through the nose to their former glory.

With all that gunk comes the feeling that it's difficult to breathe, making it feel as if it's the gunk that *causes* the difficulty in breathing.

That feeling is misleading; it's swollen nasal blood vessels that block airflow through the nose, not the gunk[590].

Colds induce two different processes to produce that gunk. One involves ramping up the mucus that usually protects the respiratory tract without drawing attention to itself. The other is a consequence of the often painfully high pressure in the nasal blood vessels, which forces blood plasma out of those capillaries to increase the volume of the gunk.

Most over-the-counter decongestants tackle both of those processes. They turn down the dial on the mucus production and at the same time, they lower the pressure in the nasal blood vessels to unblock the nasal passage and cut the flow of plasma into the gunk.

Antihistamines perform both of those functions and help with sniffles caused by conditions such as hay fever, but they're less helpful against the mechanisms that are triggered by cold viruses. A meta-analysis* found that antihistamines do soothe cold symptoms a little – but only a little[591]. They're not as effective against cold-induced congestion as decongestants sold as cold remedies, which can be divided into two categories: oral decongestants, which are swallowed, and topical decongestants, which are sprayed up the nose.

The two most widely used oral decongestants are phenylephrine and pseudoephedrine. For both of them, the evidence that they're effective against colds is somewhat patchy because of what the authors of one meta-analysis called 'considerable heterogeneity'[592] across different trials, making it difficult to combine them into a meta-analysis.

Nevertheless, the authors found that a 10-milligram dose of phenylephrine is a pretty effective decongestant. It kicked in between 15 and 30 minutes after swallowing it and kept working for three to four hours[593].

There have been fewer evaluations of pseudoephedrine, though Cardiff University's Common Cold Centre carried out one trial that found it improved nasal airflow for at least three hours after taking a dose[594].

---

* The meta-analysis encompassed trials of chlorpheniramine maleate, clemastine fumarate, brompheniramine maleate, doxylamine succinate, diphenhydramine hydrochloride, triprolidine, thonzylamine and diphenylpyraline, terfenadine, loratadine, astemizole and cetirizine, all of which are widely used antihistamines with a similar mechanism.

It would be helpful to have a head-to-head comparison between phenylephrine and pseudoephedrine but so far, nobody has done that experiment.

The evidence around the topical decongestants is more encouraging. A squirt of xylometazoline can unbung a nose for as long as ten hours and it's substantially better than its main competitor, ipratropium bromide[595]. The reason appears to be that the latter is very effective at reducing the over-production of mucus that causes sneezing and nose-blowing but it doesn't do much about the swollen blood vessels[596]. However, when the Cardiff team tried both at once, the combination worked even better than either one alone[597].

Of all the papers on all the remedies I've read for this chapter, xylometazoline stands out as the best performer. It gives a substantial amount of relief to most people who try it. Ronald Eccles, director of the Common Cold Centre that was the world's leading organisation for testing over-the-counter remedies between the 1990s and 2010s, found that topical xylometazoline improved the flow of air through a cold-afflicted nose by around seven times more than oral pseudoephedrine[598].

When I asked his view of all the cold remedies he's tested over the years, Eccles particularly recommended a puff of xylometazoline at bedtime because it's so difficult to sleep with a blocked nose and 'you wake up feeling awful in the morning'[599]. Xylometazoline can't cure a cold but because it unblocks a nose for ten hours at a time, it can be a big help in getting enough sleep while we get through the worst of it.

Cough medicines

Somewhere near the pharmacy's decongestant shelf, there is invariably another shelf loaded with different types of cough syrup. Most of them are designed around the Mary Poppins dictum that a spoonful of sugar helps the medicine go down: they use a sweet flavour to mask the taste of a pharmaceutical antitussive, as drugs that suppress coughing are called.

In 2005, a team at Pennsylvania State University looked into whether cough syrups really help what they called 'coughing children'[600] whose colds involve a cough that keeps them – and their harassed parents – up all night. They found that the most widely used antitussive, dextromethorphan, did indeed reduce the children's coughing and get them to sleep but there was a twist in the Pennsylvania tale: a spoonful of honey worked even better than a spoonful of dextromethorphan-laced cough syrup.

Sometimes a spoonful of sugar doesn't help the medicine go down so much as do the medicine's job for it. Clinical trials of antitussives rarely show them to be any more effective than placebos[601] and the Pennsylvania trial suggests why, despite the lack of evidence for antitussives, so many parents still reach for the cough syrup to suppress their children's cough. The sweet flavour alone is enough to make a child feel better.

That's because sweet flavours induce the release of opioid hormones. In a healthy person, those opioids give the sense of well-being that's familiar to anyone who enjoys a nibble of chocolate. They're the same opioids that are induced by the placebo effect of feeling we're doing something, meaning that any placebo effect is likely to be enhanced if the placebo tastes sweet[602].

However, taste sensations may offer more than a hormonally mediated placebo effect. Both sweet and bitter flavours can get the mucus flowing. More flowing mucus may feel like the last thing we need when we have a cold, until we remember that it's not the mucus that bungs up the nose but the enlarged blood vessels. Getting the mucus flowing can dislodge irritations in the throat that are triggering a cough reflex.

One way to combine sweet and bitter flavours is through the age-old cold remedy of dissolving a spoonful of honey in lemon juice and hot water. When I asked Ronald Eccles what he'd concluded from his three decades of testing pharmaceutical cold remedies, he told me, 'The first thing is hot, tasty drinks.'[603] He wasn't dismissing pharmaceutical remedies; he did, after all, lead much of the research that underpins this chapter, but his view was that hot drinks like honey and lemon – or whatever else we happen to like – make us feel better. They have the added advantage that it's impossible to overdose on a drink that has no pharmaceutical products. We can drink as much as we like, whenever we like.

Guaifenesin

In 1516, Lorenzo of Sassoferrato announced that the bark of the guaiac tree, imported from Hispaniola, was an infallible cure of syphilis[604].

He claimed to have heard about it from a mysterious spice trader he'd met in Seville, who had learned of it from the indigenous people of Hispaniola*.

Such was the terror that syphilis struck into its sufferers that Lorenzo sold his 'guiacum' all over Europe. It was completely ineffective against syphilis although it was much less harmful than the mercury compounds that were the other widely used syphilis treatment.

Guiacum arrived in Europe through Renaissance-era quackery but in the twentieth century, it was found that while the bark of the guaiac tree does not cure syphilis, it does have some medicinal properties. It functions as an expectorant, meaning that it encourages mucus to move up the respiratory tract, which has some benefit for conditions like chronic bronchitis or asthma which can be complicated by mucus blockages[605].

Guaifenesin, as the bark extract is now known, is also sold as a cold remedy, usually in combination with painkillers or as the active ingredient of cough syrup. Getting mucus moving can help to reduce coughing and a study done at the University of North Carolina found that guaifenesin did help to clear mucus out of the lower respiratory tract of people with colds[606].

However, colds don't cause much mucus to accumulate in the lower respiratory tract in the first place and if they do, it's time to consult a doctor rather than buy an over-the-counter remedy. When the researchers asked the volunteers how they felt, those who were taking guaifenesin didn't feel any better than those who were taking a placebo.

A limitation of the University of North Carolina study was that it only recruited 19 volunteers. In a much larger study, involving 366 volunteers, fully 92 per cent of the volunteers who were given guaifenesin said they would recommend the treatment they had been taking. It would be very impressive if 83 per cent of the volunteers given a placebo had not said the same thing[607].

Between those two trials, it appears that guaifenesin does help with the symptoms of colds, but not very much. It would be interesting to compare Ronald Eccles's first choice remedy of hot sweet drinks with

---

* Now divided between Haiti and the Dominican Republic.

and without guaifenesin and find out whether anyone could tell the difference.

Steaming and menthol

Like most people who are significantly debilitated by colds, I receive far more advice on how to treat them than I ask for. Perhaps the most oft-repeated advice is to steam my sinuses. If I would only fill a bowl of boiling water, I'm told, and inhale the steam with a towel draped over my head, my stuffed nose would instantly recover.

In fairness, I do get some relief when I inhale steam, but only for as long as I'm inhaling it. Spending all day under a towel is not the solution I'm looking for. But that's me. Cold remedies are nothing if not subjective and maybe I'm an outlier. Unfortunately, a delve into the research on steaming doesn't tell me whether I am or not. There have been a few trials, some of which found that steaming helps and some found it didn't. The authors of one meta-analysis called for 'more double-blind, randomised trials'[608] although they didn't explain how such a double-blind trial would be done. Volunteers tend to notice when they're leaning over a bowl and breathing steam.

Most pharmacies sell various additives that are claimed to improve the effects of steaming, perhaps the most popular of which is menthol. The pharmaceutical firm Vicks introduced menthol-based cold remedies in the 1890s and since then, tiny bottles of concoctions smelling of menthol, often mixed with eucalyptus and camphor, have become so closely identified with Vicks that long-suffering parents often call it the 'Vicks smell' as they get it out for their child's latest cold.

Menthol vapours stimulate the temperature receptors in the nose, which can give the sensation of easier breathing even though menthol makes no difference to the rate at which air flows through an inflamed nose[609]. The downside is that the time-honoured approach of rubbing it on the chest of an ailing child can cause a skin rash, albeit only in around one person in 20.

The evidence for steaming and menthol is underwhelming but they seem to work for some people and, like other remedies that do not contain a pharmaceutically active product, they have the advantage that they can be repeated as often as they need to be repeated. Nevertheless, neither look very impressive compared to ten hours of relief from a sniff of xylometazoline.

Gargling

In most countries, pharmacies sell a range of mouthwashes, but in Japan, many people treat gargling as a routine morning ablution to be done after cleaning their teeth. In 2002, Kazunari Satomura of Kyoto University led a study to find out if gargling three times per day reduced the chances of catching a cold. Over two months, he and his team found that 41 per cent of people who didn't gargle anything got a sniffle, while only 37 per cent of people who gargled an antiseptic mouthwash containing an iodine compound caught one. It wasn't a very impressive result and in statistical terms, it wasn't a meaningful difference.

Satomura's study included a third group who did better than either the mouthwash or non-gargling control group: only 30 per cent of people who gargled tap water caught colds[610].

The protective effect of tap water was statistically meaningful and probably surprising, especially as it outperformed the antiseptic mouthwash. Satomura speculated that the chlorine used to disinfect Japanese tap water may have killed cold viruses that had got into their volunteers' mouths but had yet to infect any cells. Calvin Coolidge would have approved but there's a big difference between inhaling chlorine by the lungful and the trace amounts of between 0.5 and 0.8 grams dissolved in a litre of Japanese tap water at the time.

Different countries have different chlorine concentrations in their tap water, and nobody has tested Satomura's speculation by persuading volunteers to gargle water with different chlorine concentrations. Moreover, Satomura's speculation rather begs the question of why chlorine inactivated viruses so much more effectively than iodine.

There is an alternative explanation, which is that the tap water inactivated viruses not with dissolved chlorine but through what's called osmotic stress. Human viruses evolve inside the human body, in which the fluids contain a certain amount of mineral salts. The cells of the human body, both human and bacterial, are mostly made up of fluids containing salt at the same concentration and don't react well to being placed in fluid with a higher or lower concentration of dissolved salt. Many human viruses are equally sensitive to salt concentrations so it's possible that regularly flushing the mouth with pure water lowers the salt concentration enough to inactivate at least some of them.

Some mouthwashes are designed to kill viruses through osmotic stress, although such 'hypertonic mouthwashes' do so by flushing the mouth with too much rather than too little salt for viruses to handle. Hypertonic mouthwashes have not been evaluated very often, probably because they don't contain any pharmaceutical products. However, Sandeep Ramalingam of Edinburgh University led a study that showed that a combination of nasal irrigation and gargling with sea salt dissolved at between 1.5 and 3 grams per litre shortened colds by around a fifth[611].

Neither Satomura nor Ramalingam claimed to have discovered a miracle cure but their studies on tap water, with or without salt dissolved in it, returned much more impressive results than most evaluations of branded mouthwashes, none of which have proved consistently effective against colds[612].

Green tea

In 2016, a meta-analysis by researchers at Japan's University of Shizuoka combined the results of five different studies and found that regularly gargling green tea prevented between one in five and one in ten influenza infections[613]. Their result comes with a few caveats. For one thing, influenza viruses only cause a small minority of colds. By far the most prolific cold viruses are the rhinoviruses, which are much more robust than the influenza viruses.

Another concern is common to any study on the medicinal properties of a plant: the biochemical content of a species of most plants varies widely. The presence or concentration of any medicinal substance depends on factors like the variety of the plant and the soil it grows in. With green tea, any antiviral properties may also be affected by the way the tea leaves are processed before their biochemicals are dissolved and poured into someone's throat. That processing is, after all, the main difference between green tea and dark teas that have not been assessed as influenza preventatives.

Other types of tea may be just as effective although, even if they are, we shouldn't be too impressed. The protective effect of green tea gargling is in the same range that Satomura's team found with Japanese tap water although, like Japanese tap water, green tea does outperform branded antiseptic mouthwashes.

One thing we do know is that green tea contains a class of protein molecules called catechins which interfere with the process by which some viruses enter cells. Simply drinking green tea or taking catechins in capsules has often been reported as being as protective as gargling[614]. It's no miracle preventative but it's good news for anyone who enjoys a cup or two of green tea every day.

Nasal sprays

Many healthcare professionals recommend regular nasal irrigation, involving washing the nose out with saline solution, to get rid of irritants like dust and pollen. Whether or not it works is an open question. The last major meta-analysis of irrigation only found two fairly small-scale studies and the authors didn't rate either of them as being very well conducted[615].

Dealing with cold viruses, which actively infect cells, is very different to dealing with irritants that don't get past the lining of the mucosal tract. However, the evidence for using nasal sprays against viruses is no better than the evidence for using them against irritants[616]. None of that is to say that nasal sprays *don't* work against either irritants or viruses. It just means that there's frustratingly little evidence on whether they work or not. A more significant finding was that although saline solution contains no pharmaceuticals, nearly a quarter of people who use it regularly report some sort of unpleasant adverse event like headaches or nosebleeds from using it[617].

That still means that three-quarters of the people who used it had no problem with saline. The best conclusion to be drawn is that it's probably worth trying for things like hay fever and perennial rhinitis that cause a persistent sniffle. The unlucky one in four who suffer from adverse events can simply stop. Using it to prevent colds, however, is at best a long shot.

Those studies investigated isotonic saline, containing mineral salts dissolved to the same concentration as in the human body. That concentration is substantially lower than the hypertonic saline solution that Sandeep Ramalingam found shortened colds, but as he used that hypertonic saline for both gargling and washing, it's impossible to know whether it was either or both of those that were shortening his volunteers' colds.

In recent years, a new generation of nasal sprays has appeared that contain plant-derived polysaccharide molecules specifically intended to target cold viruses. Such polysaccharides have been shown to entrap viruses, as well as various other irritant particles, in laboratory culture.

Laboratory experiments only go so far. A lot of pharmaceuticals that look promising in the laboratory turn out not to do anything in the respiratory tract but at least one of those polysaccharides looks promising.

Carrageenan is extracted from a red seaweed called Irish moss. It's widely used to thicken products like sauces, ice cream and toothpaste, which doesn't prove it has any medicinal effects, but it does give it a tried and tested safety record. At the time of writing, three trials have assessed carrageenan nasal sprays as cold remedies. In 2021, a meta-analysis combined the results of two of them and found that the carrageenan spray halved the likelihood of a cold persisting for more than two weeks when compared to a saline spray[618].

A third study, done since that meta-analysis was published, was less encouraging. Volunteers using carrageenan found that their colds didn't clear up any faster than volunteers who were using simple saline solution but were more likely to get persistent headaches[619]. However, that more recent study included very few volunteers whose colds lasted as long as two weeks, suggesting that carrageenan may stop a cold from dragging on but won't make any difference to a cold that was always going to clear up in a week.

Carrageenan is not the only polysaccharide that can be squirted up the nose. Some nasal sprays contain a plant polysaccharide called hydroxypropyl methylcellulose – biochemists love their polysyllables as much as the most classically minded virologist – which is also marketed as a cold remedy. Unfortunately, I can't say whether it's any good because at the time of writing, no reliable evaluation of hydroxypropyl methylcellulose has been published[620].

I asked Ronald Eccles, who led one of the carrageenan trials, and he agreed that the evidence is equivocal.

'It's not definitive at all,' he told me, 'but I think it is useful'.

His view is that carrageenan-fortified nasal sprays won't do any harm – unless you're one of the unfortunates in whom it causes a headache, in which case they're best avoided – and they might do some good, so why not use them? Eccles even uses carrageenan as a preventative, for which it's never been tested. He uses a shot or two to protect himself when he places himself in a situation where there's a high risk of catching a cold, such as when he's taking a commercial flight[621], taking the view that it doesn't hurt and it might help.

In praise of the placebo effect

If you've got this far and you're seething that I haven't mentioned your favourite cold remedy, it's probably because I found no good evidence that it does anything. It may cheer you up to know that such seething places you in the company of double Nobel laureate Linus Pauling.

It's the sort of seething that led him to introduce cold research to meta-analysis. He received – and deserved – a considerable amount of criticism for his late-life fixation with vitamin C but he gave cold research a shove in the right direction, even as Pauling himself was careering off on the wrong one. However, he only resorted to meta-analysis because the effect of vitamin C was weak. If vitamin C supplementation consistently protected against colds, it would have been clear in every published study and Pauling would not have needed to introduce meta-analysis to cold research.

As with vitamin C, so with most of the cold remedies described in this chapter. Meta-analysis is a very powerful tool for reconciling contradictory results but if there's a contradiction to be reconciled, it's probably because the effect is not powerful enough to show up in every study. There is reasonable consistency with the more powerful analgesics like paracetamol and ibuprofen and with the decongestant xylometazoline. For most other products, different studies return different results. Moreover, every trial of a cold remedy reports a very wide range of effects from different volunteers. Most remedies appear to work in some people but not in others.

The inconsistencies that drive cold researchers to meta-analysis are a product of the inconsistencies in colds. When a cold can be caused by any of over 200 viruses and when different peoples' immune systems respond in different ways depending on how many times they've been infected by similar viruses, how stressed they are and, quite possibly, whether they've had a good soaking, it's hardly surprising that medications that work on one cold or in one person don't work on another cold or another person.

There is another reason that's rooted in the fact that colds are inherently subjective. An inflamed upper respiratory tract is as unpleasant as it feels. No more, no less. It follows that a remedy that makes a cold less unpleasant is working even if it has no measurable physiological effect. Even if an improvement is driven by a placebo effect, it's still an effective improvement.

The power of the placebo was shown in the guaifenesin trial in which four out of five volunteers who took a placebo said they would recommend it as a cold treatment[622].

Nobody is more familiar with the placebo effect than Ronald Eccles, who has spent decades giving volunteers placebos to test cold remedies at the same time as researching the placebo effect itself[623]. If anyone can be immune to the placebo effect, it would be Eccles, but he said to me, 'I'm as subject to a placebo effect just as much as everyone else.'[624]

So, if I haven't mentioned a favourite remedy of yours, there's no need to feel frustrated. Maybe the remedy works for you and a few other people, but not enough people to sway the result of a clinical trial. Maybe you're enjoying the benefits of a placebo effect, in which case there's no reason for hard data to spoil a good thing.

When you find something that works, you have a plan for dealing with your next cold. Take your remedy, pour yourself one of those tasty drinks that Eccles recommends and try to get through the next few days as painlessly as possible.

Chapter 13

Personal protection

If we have to spend much of our time inhabiting an environment that could have been designed to transmit colds which, as Chapter 5 described, many of us do, it raises the question of how we can protect ourselves while we inhabit it. Vaccination offers an answer, but only a partial one. The available vaccines only cover a few of the more than 200 cold viruses in circulation. Against the others, our best bet is to keep them away from our mouths and noses. They can't do us any harm if they can't infect us, and there are two simple measures we can take to stop them: wash our hands and wear a facemask.

There's nothing new about personal hygiene or trying to purify air before breathing it in. In medieval Europe, where it was believed that disease was caused by foul smells, people often tried to protect themselves by breathing through scented handkerchiefs. It wouldn't have done any good against airborne microbes but before sewers were closed and placed underground, it probably had its advantages.

The man who brought face coverings and hand hygiene into the microbial era was an Austrian surgeon called Johann Mikulicz-Radecki. The seeds for Mikulicz-Radecki's war on germs were sown in 1879, when he travelled to London and met the pioneering English surgeon, Joseph Lister.

At the time, the microbial theory of disease was slowly gaining acceptance due in no small part to Lister's efforts. For years, Lister had been disinfecting wounds, bandages and surgeons' hands with carbolic acid. His colleagues may have thought he was eccentric, but they couldn't argue with his results. At a time when a patient undergoing surgery had a fifty-fifty chance of dying of 'putrefaction', the almost complete absence of putrefaction among Lister's patients was impossible to ignore[625].

Lister's success made his department at London's King's College Hospital into one of Europe's major medical centres. He often hosted young surgeons like Mikulicz-Radecki on the study tours that were then an essential step toward building an eminent career.

When Mikulicz-Radecki returned to Vienna, he incorporated disinfection into his own practice, but he was unimpressed with carbolic acid. It was nasty stuff that tended to kill the tissue around a wound or incision, slowing the healing process. It wasn't much fun to use either, often leaving surgeons with the skin sloughing off their sore hands. Mikulicz-Radecki valued his hands; he was such an accomplished musician that he partnered with Johannes Brahms to premiere the composer's piano duets[626].

Mikulicz-Radecki found that iodine was just as effective against microbes as carbolic acid but much gentler on both the patient and the surgeon, which made the latter far more likely to use it.

Disinfection with iodine was one of Mikulicz-Radecki's many innovations. He went on to invent many items of surgical paraphernalia that still bear his name and, after he moved to Germany to take up the post of head of surgery at the University of Breslau*, he invented the precursor to the modern facemask. By then, middle age and promotion had turned Mikulicz-Radecki into something of a martinet, who disapproved of any exchange between his subordinates that went beyond the purely functional. Junior surgeons may have continued to greet each other with handshakes and call each other *Herr Kollege*, but only when Mikulicz-Radecki was not watching[627].

Mikulicz-Radecki knew that everybody breathes out tiny droplets laden with microbes, and he knew that speaking projects those droplets through the air. He introduced a system of hand gestures that allowed surgeons to communicate without speaking, but it only worked up to a point. However much time he spent on refining the hand signals, something always seemed to come up that his system didn't cover.

He took to tying strips of gauze across his mouth and beard – then very much in fashion among the medical men of Europe – to contain any microbes he might project[628]. In 1897, he mentioned his *mundbinde*[629], meaning 'mouth bandage', in a description of the measures he

---

* Now the University of Wrocław in Poland.

used to protect his patients from infection. It was the first description of what we now call a surgical facemask and, alongside his innovations in hand hygiene, it made him the man behind today's methods of personal protection against the common cold.

Hand hygiene

The principle behind hand hygiene is straightforward: a virus-laden droplet can settle on a surface and if we touch that surface, we may pick up the virus on our hands. If we then touch our face, we may infect ourselves by letting the virus into our nose or mouth.

Soap dissolves the protein capsid that protects the virus's genetic material, making simple handwashing very effective against any viruses we might have picked up. However, soap requires running water and when we're coming out of a crowded place where we're likely to have encountered cold viruses, say a train carriage or a cinema, there may not be a sink available.

Any pharmacy will stock a range of ethanol-based hand sanitisers, which don't need water because ethanol evaporates in less than a minute. The range of brands can be bewildering but any sanitiser containing 50 per cent ethanol or more will inactivate most viruses within seconds, including the influenza[630] and coronavirus[631] families that include the worst of the cold viruses.

Most of us regard hand hygiene as a matter of common sense but then, common sense is often less than common. Several different studies have shown that reminding people to wash or sanitise their hands regularly substantially reduces the number of colds people catch, and even more substantially reduces the number of stomach bugs they pick up[632]. If nothing else, those studies show that a lot of people need reminding. We touch a lot of things in a normal day so it's worth taking the time to wash off any nasties we might have picked up.

Hand hygiene matters, but it only goes so far. We were painfully reminded of that in the early stages of the COVID-19 pandemic, when a lot of the public health messaging emphasised hand hygiene as the single most important precaution[633]. However, no amount of hand hygiene is going to help if the virus is floating around in the air.

Nevertheless, hand hygiene did have a role to play; SARS-CoV-2 can survive on surfaces and on skin[634], although it's not entirely clear for how long. The question is why the WHO, and the many national public health agencies that followed its lead, continued to downplay the importance of airborne transmission long after it was clear that it flew in the face of the evidence. The century-long controversy about airborne transmission described in Chapter 3 played its part, though public health experts may also have been misled by so much common cold research having been focused on the rhinovirus family.

The rhinoviruses cause between a third and a half of all colds[635], which is more than any other type of cold virus. Rhinoviruses behave like a regular cold virus when they infect the respiratory tract but outside the body, they are considerably tougher than any coronavirus. That 50 per cent ethanol hand sanitiser that demolishes SARS-CoV-2 doesn't do much to a rhinovirus. One study showed rhinoviruses shrugging off a much stronger 80 per cent ethanol solution[636] and they've even been shown to survive treatment with hospital-grade disinfectants[637]. That resilience is likely to make rhinoviruses far more likely to survive the transition from a surface to a hand to a face than a coronavirus, or indeed any other cold virus.

All too often, results obtained from one cold virus are extrapolated to cold viruses in general, which can be misleading when different virus families have such dramatically different characteristics. In this case, it may have led to the assumption that all cold viruses are as capable of being transmitted by our hands as the highly resilient rhinoviruses.

From very early in the COVID-19 pandemic, it was evident that like most cold viruses, SARS-CoV-2 mostly transmits through airborne droplets[638]. More than a century earlier, Mikulicz-Radecki had pointed the way to a solution, although the modern facemask has come a very long way from his *mundbinde*.

The making of the modern facemask

The first improvements to the *mundbinde* were made by Mikulicz-Radecki's assistant, Wilhelm Hübener, who inveigled his colleagues into experiments with various mask designs. He must have strained their tolerance because those experiments were deeply unpleasant. They involved swallowing a mouthful of bacteria cultured in brine and decomposing herring, then speaking through a mask to see which design was most effective at stopping them from spraying bacteria onto a plate of nutrient agar[639].

Thanks to his colleagues' forbearance, Hübener hit on a design involving multiple layers of gauze that covered the nose as well as the mouth[640]. His facemasks went a long way toward protecting patients from doctors, but it would be another decade before facemasks were adapted to protect their wearers.

In 1910, a plague was sweeping through the north-eastern Chinese province of Manchuria. The man tasked with getting it under control was Wu Lien-teh of the Imperial Army Medical College in Tianjin. Wu was well versed in the microbial theory of disease, having followed a Cambridge medical degree with research under one of the theory's pioneers, Élie Metchnikoff at Paris's Institut Pasteur[641].

In Manchuria, Wu quickly established that the epidemic was being caused by the same bacterium that causes bubonic plague. *Yersinia pestis*, as the plague bacterium is now known, was a familiar foe but it was doing something unfamiliar. Bubonic plague is carried by fleas, but fleas don't get the bacteria into human lungs, which was what Wu was seeing in Manchuria. The plague bacterium, he realised, was airborne. Wu was dealing with the first recorded epidemic of what is now called pneumonic plague[642].

A novel problem inspired Wu to a novel solution. He issued medical staff with a modification of the surgical facemask, involving layers of cotton and gauze tied on tightly enough that any air breathed in was filtered through the mask[643].

The epidemic lasted through the winter of 1910 to 1911 and while the quarantining and masking imposed by Wu probably wasn't the reason it ended, it certainly saved a lot of people from being infected. It also introduced the idea of a facemask intended to protect its wearer rather than to protect someone else from its wearer.

The idea returned to the fore during the 1918 influenza pandemic. In Britain, Harrods and Selfridges reported a brisk trade in facemasks, although some newspaper reports suggest that most people saw them as something of an oddity. One woman who wore a mask in the London suburb of Tooting was reportedly followed by children for over a mile[644], suggesting that masks were not widely used outside the fashionable set who shopped in London's West End.

In the USA, several city authorities introduced compulsory masking which, in a pattern that would be repeated during the COVID-19 pandemic, was initially popular before being met with a backlash. In San

Francisco, people who refused to wear a mask were initially derided as 'mask-slackers' and reviled as thoroughly as draft-dodgers. By January 1919, the mood had changed. The Anti-Mask League had formed to petition against mask mandates, even as the pandemic's third wave arrived[645].

After the pandemic waned, facemasks were rarely seen in public, although they became progressively more widely adopted by hospital staff. By the 1930s, they were a standard piece of what we now call personal protective equipment or PPE[646].

The facemasks used during the 1918 pandemic and subsequently were developed directly from Wu's design. They were made of plant-based fibres, usually cotton, arranged into six layers which made them thicker and probably less comfortable than the facemasks in use today. The next step toward today's facemasks was driven by the mid-twentieth-century boom in plastics.

In 1958, Sara Little Turnbull got a call from 3M, one of the USA's largest manufacturing conglomerates. They had developed a line of stiffened ribbons made from a technique called meltblowing, in which molten polypropylene – a type of plastic – is forced through tiny holes to form fibres. The executives of 3M believed that meltblowing might have applications that went beyond gift wrapping and they wanted Turnbull, a former décor editor for *House Beautiful* magazine and now an independent design consultant, to consider what other meltblown products might turn a profit.

Turnbull did not disappoint. In a presentation titled *Why?* she suggested more than a hundred different products to 3M's executives. Her audience was all male, which may or may not be why the product they commissioned her to develop was a moulded bra cup. While she was working on it, Turnbull's life took a dark turn as three people close to her died in two years[647]. She spent a lot of time visiting hospitals and when she saw the facemasks that the nurses of the time were wearing, she had another idea.

Turnbull reasoned that if meltblowing could produce a structure that fitted a breast, it could produce a structure that fitted a face. She modified her bra cup design to invent the first respirator mask: a facemask made of plastic fibres that forms a partial seal, filtering not only the air breathed out but also the air breathed in through those fibres.

Turnbull's meltblown masks filtered air in much the same way as Mikulicz-Radecki's *mundbinde*: the fibres broke up the air currents caused by breathing and as the air was deflected in different directions, dust particles and droplets collided with the fibres. We can imagine the process as similar to what happens when a tree branch falls into a stream. The water keeps flowing over, under and around it but twigs and leaves get tangled up in it.

However, the physical force that keeps dust particles and droplets stuck to mask fibres is very different to the forces that tangle debris in a submerged tree branch. At the tiny level of airborne droplets and polypropylene fibres, solid objects that collide tend to stay stuck together[648]. If you want to see those forces in action, do some dusting. It's the same forces that stick dust particles to the duster's fibres and lift them off the surface you're dusting. Those forces are not particularly strong; dust particles are about the largest objects that they can operate on. That's why we can lean against a brick wall without worrying that we'll get stuck to it. Those same forces are in operation, trying to stick us to that wall, but they're so weak that we don't notice them.

So weak, it turned out, that they didn't capture enough virus-laden droplets passing through a facemask to protect its wearer from infection. Turnbull's masks filtered dust fairly well and they were widely adopted to protect construction and mining workers, very few of whom knew they were wearing a modified bra cup on their face, but saw little uptake among the doctors and nurses for whom Turnbull intended them.

The man who changed that was Peter Tsai, a material scientist at the University of Tennessee. In the early 1990s, Tsai worked out a technique for adding an electrostatic charge to the fibres. Electrostatic attraction is the force that makes a balloon stick to things after it's been rubbed on a T-shirt and, being a force that's able to act on things large enough for us to see, it's considerably stronger than the forces that stick dust to a duster. Tsai's treatment made meltblown facemasks ten times more effective at filtering the air passing through them[649] and opened the door to a practical, affordable facemask that protected the wearer from airborne infection.

Face masking in the twenty-first century

By the dawn of the twenty-first century, three facemask types were available: respirator masks, surgical masks and cloth masks.

The first choice for personal protection was and remains the respirator mask, which emerged out of a century of development driven by Wu, Turnbull and Tsai. Polypropylene is now one of several synthetic materials used to make facemasks, but they all work in the same way: air passes through a maze of electrostatically charged fibres that trap dust, droplets and microbes[650]. In Britain and the European Union, respirator masks are required to pass the 'FFP2' standard by demonstrating that they can filter out 94 per cent of all airborne particles, although they are often called 'N95', referring to the equivalent standard in the USA*.

Surgical masks are made from the same materials as respirator masks, but they do not seal around the face. They're very effective at filtering microbe-laden droplets out of an exhalation but, as the illustration on page 249 shows, so not all the air the wearer inhales is filtered through the mask's mesh.

Although purpose-made facemasks are widely available, cloth masks never went away. They've become a familiar sight since 2020, when the COVID-19 pandemic drove a global shortage of respirator and surgical masks. Cloth masks are cheap, can be made by anyone with basic sewing skills, can be decorated with pretty pictures, and they filter at least some droplets out of the air passing through them[651]. How many droplets depends on the cloth they are made from. Fabrics that stretch make poor filters because the stretching expands the gaps between the fibres. The best options are cotton, muslin and flannel and if in doubt, a mask from at least three layers of bath or tea towel can have gaps between the threads[652] as narrow as those of a surgical or respirator mask[653].

However, there is a reason why surgical and respirator facemasks are not made from tea towels. However tightly packed a cloth mask's fibres are, they do not carry an electrostatic charge.

---

* For any facemask nerds, FFP stands for filtering facepiece and there are three different types. The FFP1 is a dust mask that is not particularly effective against microbes, while the FFP3 is used for situations where there is a risk from very fine, particularly hazardous dust particles such as asbestos. Some FFP2 masks have an exhalation valve, in which case exhaled air is not filtered. In this book, a respirator mask means FFP2 or equivalent without an exhalation valve unless otherwise specified. The standards are based on a set of filtration tests that differ slightly between different countries but in practice, the FFP2 standard is equivalent to the USA's N95, the Chinese KN95, the New Zealand and Australian P2, the Japanese DS2, the South Korean KF94 and the Brazilian PFF2.

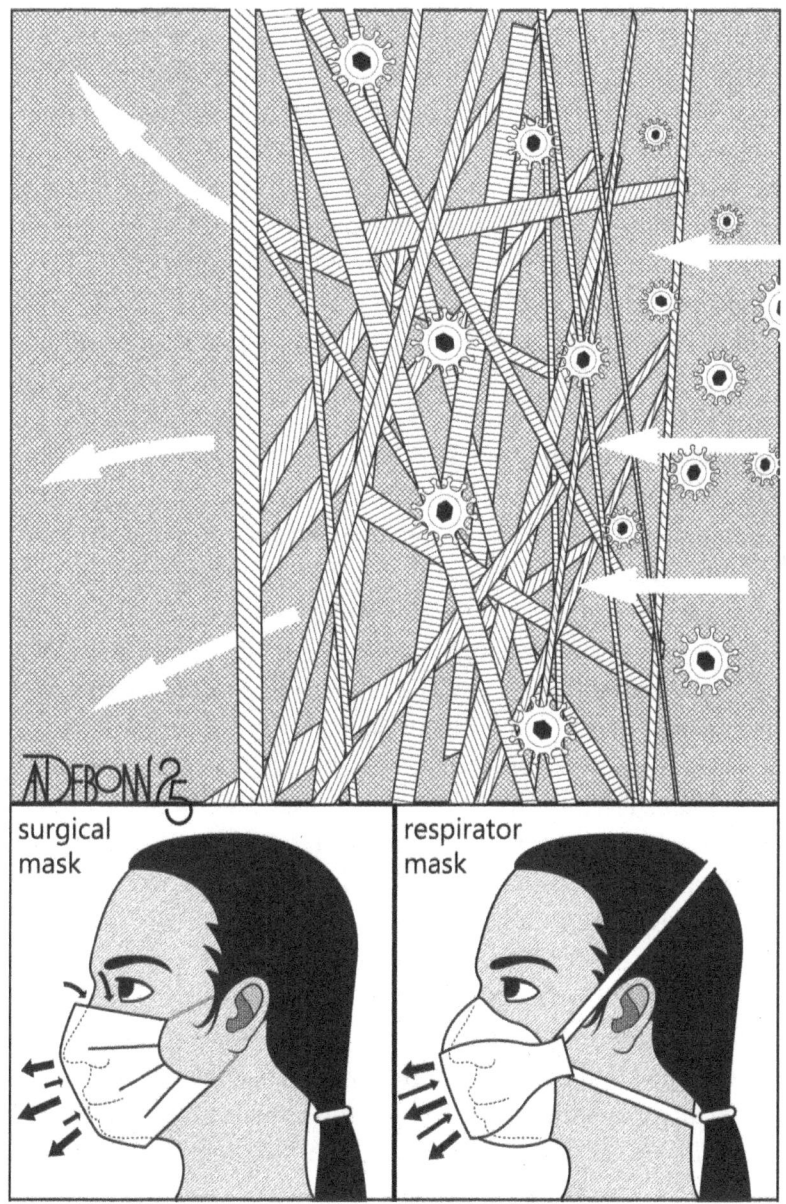

The top panel shows the mesh of facemask fibres filtering airborne viruses out of air flowing through the mask material. The virus capsids and fibres are not to scale; the diameter of the fibres is typically 5–50 times the diameter of a virus capsid. The lower two panels show the airflow through a surgical and respirator mask. Both masks filter the exhaled breath but only the respirator mask filters all of the inhaled breath. The surgical mask, which is not sealed to the face, allows air to be drawn around the edges of the mask material.

In 2000, the respirator mask was still a recent innovation and, like the surgical mask, it was used mostly to stop the spread of infection in hospitals and clinics. That changed in 2003, with the first airborne epidemic of the millennium. The SARS epidemic* triggered a raft of public health measures in the east and south-east Asian countries it swept through, one of which was to bring facemasks out of clinics and hospitals and into trains, shops and anywhere else where a lot of possibly infectious people gathered together.

In the aftermath of SARS, epidemiologists put a lot of effort into evaluating the various public health measures that had been deployed against it. If they could work out what had helped and what hadn't, public health authorities would be able to deploy a more effective response to the next airborne epidemic.

A review of public health measures used in Beijing showed that people who wore a mask when they went out were indeed less likely to catch SARS than those who didn't[654] but they also noted that if someone was cautious enough to wear a mask, they were probably taking other precautions as well. The authors couldn't conclude that the protection came from the mask and not from whatever those other precautions were, illustrating a problem that comes up repeatedly when trying to learn from responses to SARS: an emergency epidemic response involves deploying a lot of measures. If the response is followed by an amelioration of the epidemic, it's difficult to work out which measures were doing the ameliorating and which did nothing but happened to be deployed at the same time.

During the 2009 'swine flu' influenza pandemic, the facemask attracted the attention of public health professionals around the world. It soon emerged that the new strain of influenza was no more dangerous than regular seasonal influenza, but it did spur more detailed research into whether a facemask was indeed a viable approach to personal protection.

The value of the facemask

Two questions came to the fore. The first was how long a facemask would remain effective. While respirator masks only qualify as such

---

* See Chapter 9 for a description of the SARS epidemic.

after passing stringent filtration tests, those tests are done on masks taken straight out of the box. Nobody had checked how long a mask continues to meet those standards.

The second question was simpler but fundamental: did filtering inhalations and exhalations translate into protection from airborne infection?

While facemasks were mostly used as medical PPE, the first question attracted little attention. Doctors and nurses tended to wear their masks while treating infectious patients and then throw them away. The COVID-19 pandemic changed that almost overnight. Suddenly, everyone wanted a respirator mask to go to the supermarket. The manufacturers couldn't keep up with the demand, and we all had to make whatever masks we could get a hold of last as long as we could.

The task of finding out how long that was landed in the lap of Hao Chen of the Oak Ridge Institute for Science Education in the USA, who answered it by persuading volunteers to wear the same respirator mask for eight hours every day. He found that after five days, amounting to 40 hours of continuous use, three of the four brands of mask he tested were filtering just as effectively as before they had come anywhere near anyone's face[655]. Even the fourth, manufactured by Honeywell, fell only slightly below the filtration efficiency standard that had got it certified. It probably helped that the volunteers were taught how to look after their masks when they weren't wearing them: they kept them in paper or mesh bags, which protected them from chafing and gave them a better chance to dry out than a plastic bag would, and they kept them flat to avoid crumpling the tangle of meltblown fibres.

Chen showed that respirator masks are so much more robust than had been suspected that, with the exception of the Honeywell product, he didn't even get as far as finding out how long a respirator mask lasts before it starts to deteriorate. The other three* might have kept on filtering through another month of daily wear or they might have gone sharply downhill after another day. However, there's another reason why we're going to want to change masks in less than 40 hours.

---

* The other three respirator masks were produced by Shenzhen BAK Medical Technology (KN95 certification), KM Corporation (KF95 certified) and 3M (N95 certified).

A smaller study, carried out at the same time as Chen's, agreed that respirator masks retained their filtration efficiency after 40 hours, but stopped there because the masks became 'malodorous'[656]. That should not have surprised anyone; 40 hours is a long time to wear anything next to the skin. The smell points to a limitation of both surgical and respirator masks, which is that washing them strips them of their electrostatic charge[657], effectively turning them into another type of cloth mask.

Taken together, the various studies show that respirator masks filter air for at least 40 hours if they're treated right.

Showing that masks designed to filter air do in fact filter air in laboratory tests is one thing. Showing that facemasks protect us in the real world, amid all the messiness of real life that experiments are brought into laboratories to avoid, is another. Before recommending mass mask-wearing, public health policymakers wanted to know if all that filtering translates into protection from airborne infections in the real world.

Face masks in the wild

Along with a range of facemasks, the twenty-first century brought two good reasons to use them: first came the rise of a dangerous avian influenza, or 'bird flu', that was identified in Hong Kong in 1997[658], and then came the SARS epidemic of 2002 to 2004. Face masks blocked the transmission of such airborne viruses in laboratory studies, but real-world settings are much messier than laboratories.

To prepare for future epidemics, policymakers needed to know whether facemasks could be used to protect people against them. The catch was that they needed to know that *before* there was an epidemic to test facemasks against. Cold viruses offered a solution. They're transmitted in the same way as avian influenza and SARS, there are cold viruses that are closely related to both and colds are always there to run a test against.

Many of those trials produced frustratingly inconsistent results, probably because the trials themselves were inconsistent[659]. Some studies used surgical masks, some used respirator masks and some were reported in papers that did not specify which. Some trials involved instructing volunteers to wear facemasks at the same time as they instructed them in improved hand hygiene techniques, making it impossible to tell whether a reduction in infection was down to volunteers wearing masks or washing their hands more often.

Perhaps the most important problem was revealed by a study led by Raina MacIntyre of the University of New South Wales. Her starting point was that, as Owen Lidwell showed back in the 1950s*, it's usually children who bring colds into a household. She reasoned that if she could identify households in which children brought home colds, she could divide the families into two groups: those whom she asked to wear respirator masks and those whom she didn't. It was a simple study design that, MacIntyre reasoned, would show if respirator masks could protect family members from a child's cold virus.

When the study finished, MacIntyre found there was no difference between the two groups[660]. However, she didn't conclude that respirator masks were ineffective because she'd done something that, remarkably, no previous study of masking had done: she checked to see whether the people she asked to wear a mask actually wore a mask.

They didn't.

Some people complained that the masks were uncomfortable, some said masks upset the children and many simply said they forgot about them. It flagged an important issue that many facemask studies overlook: volunteers who are asked to wear facemasks do not always wear them. Any assessment of whether facemasking is protective must include an assessment of how much time the facemasks spend on faces instead of in pockets.

Because she asked that simple question, MacIntyre knew she had no useful information about whether facemasks protected from colds but she could conclude that people weren't going to wear facemasks around the family home – at least not without something much more serious than a cold to worry about. MacIntyre's thoughts turned to hospitals where, as she put it to me, 'You're surrounded by sick people, you've got other health care workers who are sick, you're getting continuously and repeatedly exposed to respiratory infections.'[661]

However, she quickly found that Australian doctors and nurses were no more willing to wear facemasks than the families of children

---

* Chapter 5 describes Lidwell's studies in rural and urban communities.

with colds. She discussed the problem with colleagues in China who suggested that Chinese healthcare workers might be more amenable, not least because they'd had to deal with the SARS epidemic that had not seriously affected Australia.

In the winter of 2008–2009, MacIntyre and her colleagues assessed both surgical and respirator masks across 15 Beijing hospitals. They found that staff in hospitals issued with respirator masks were far less likely to pick up cold viruses, as well as other airborne infections, than staff issued with surgical masks or in hospitals where masking was not a requirement[662].

Respirator masks protected the people wearing them.

However, MacIntyre's result didn't send doctors and nurses to the hospital store to demand more masks. Nurses issued with respirator masks were more likely to complain of discomfort than those issued with surgical masks. They were also more likely to say their masks interfered with communicating with patients and were more likely to 'forget' to wear their masks.

It's easy to be dismissive of their concerns, especially when there's no reason why a respirator mask would make communication any more difficult than a surgical mask. However, it may be relevant that the nurses in question were Chinese and more than half were women. They may have fallen foul of a pernicious legacy that lingers from the way the original respirator masks were designed.

How well those nurses' respirator masks protected them would depend on how well the masks fitted their faces. The better the fit, the better the seal around the edges and the more of the air that they breathed would pass through the mask's fibres.

The original respirator masks were designed around measurements taken by the United States Air Force in the 1960s. Consequently, early respirator masks were designed for faces that were white, male and clean-shaven. Such masks were too large to fit most women, although that issue has been at least partly rectified by the much wider range of sizes now available.

A more pervasive problem is that respirator masks designed for Caucasian faces tend not to seal around African or East Asian faces particularly well[663]. The situation has improved since MacIntyre's Beijing study because respirator masks designed for the East Asian market, which look more like a tent than a modified bra cup, are now widely available around the world. At the time of writing, nobody has checked whether they fit African faces any better than masks designed for the 1960s US Air Force. Nor is there a respirator mask that forms a seal around facial hair.

A further problem was that one in every five of MacIntyre's interviewees said that a respirator mask made it difficult for them to breathe which, taken at face value, appears unlikely. Even if the masks did not fit well, that would lead to air getting past the filter through gaps between the mask and the skin rather than blocking air from getting through at all. Moreover, many different studies have shown that wearing a respirator mask does not affect breathing and heart rate, even when those studies have involved giving volunteers a thorough workout on treadmills or exercise bikes[664]. Nor would we expect them to; an oxygen molecule's diameter is about a hundredth of that of a rhinovirus. It can easily slide between a mask's fibres.

However, what those doctors and nurses were saying was that the mask was so uncomfortable that they *felt* they were having trouble breathing while they were wearing it. They weren't about to pass out, but mask designers would be well-advised to pay attention to that feeling. If respirator masks are so uncomfortable that they feel like they're interfering with the wearer's breathing, then facemask development is still a work in progress.

One obvious improvement would be to address an enduring mystery of facemask design, which is why so many brands are fastened with elastic loops around the ears. A loop around the back of the head forms a better seal around the face[665] and doesn't feel like it's trying to pull the wearer's ears off after a few hours.

However, even a poorly fitting mask will still filter far more inhaled air than no mask at all.

Face masks vs COVID-19

MacIntyre's Beijing trial was the first to show that respirator masks work and later studies backed her up. A recent meta-analysis showed that even compared to a surgical mask, a respirator mask cuts the risk of airborne infection by between one-third and two-thirds[666].

When the COVID-19 pandemic broke upon the world, that message had not been particularly widely received. The World Health

Organization was still in denial about airborne transmission and when public health bodies began to advocate facemasks, they were usually talking about surgical rather than respirator masks.

In some parts of the world, facemasks became as controversial as in 1919 San Francisco. For epidemiologists, it was frustrating to see people railing against a cheap and simple public health measure, but differing patterns of facemask use did offer an opportunity to thoroughly evaluate facemask effectiveness. Some of the best data has emerged from American schools, where each school's management decided whether or not to mandate facemasking in the classroom.

Earlier studies, like MacIntyre's, had been limited because researchers could not force anyone to wear facemasks. They could only ask volunteers to wear them and check to see whether or not they did so. School rules are a different matter; if a head teacher says everyone needs to wear facemasks, teachers will ensure that everyone wears facemasks.

The results were consistent across multiple studies that compared schools with and without facemask mandates in the same districts. Separate assessments in Arizona[667], Florida[668], Georgia[669] and Massachusetts[670] showed that facemask mandates reduced the risk of COVID-19 for pupils and teachers by between one-third and nine-tenths. That's a very wide variation, but school facemask mandates can only control the risk at school. They reveal nothing about who might or might not be wearing masks in shops, public transport or out-of-school clubs.

The masks worn in those schools were surgical rather than respirator, and the dramatic effect on COVID-19 infections shows the value of masks as a community intervention. Surgical masks interrupt airborne infections if everybody is wearing them, which is a valuable lesson for a future airborne pandemic. However, the school studies don't tell us anything about how useful a surgical mask is when few other people are wearing them. If most people are not wearing facemasks, a respirator mask is a much more effective way to protect ourselves than a surgical mask.

In 2023, the British Royal Society commissioned a review of all studies published to date on the effect of facemasks on COVID-19, which led to the conclusion that 'wearing masks, wearing higher quality masks (respirators), and mask mandates generally reduced the transmission of SARS-CoV-2 infection'[671].

The statement is laden with the hedging that pervades academic writing, but the message is clear. Respirator masks work. They're not impenetrable but they substantially reduce the risk of being infected with SARS-CoV-2 or any other airborne virus.

To mask or not to mask

There's strong evidence that we can protect ourselves from colds with hand hygiene and facemasks but, assuming we're not going to live the rest of our lives standing over a sink washing our hands while wearing a respirator mask, we're going to have to make some decisions about how to incorporate that evidence into our daily lives. Those decisions are likely to be based on a combination of what the risk of infection is, what the risk to ourselves is and what we can reasonably do about it.

The risk depends on how likely it is that the air we're breathing contains a cold virus, which in turn depends on how many people that air has passed through recently and how likely it is that one of those people has breathed some virus into it. If we're outdoors, not much of that air has recently been in and out of anyone's airway so it's unlikely to contain much virus. Somewhere with a lot of people passing through, like a train carriage or a cinema, will contain the respiratory contributions of a lot more people. That contribution will linger after the people who contributed it have left. Some of it will be in droplets small enough to float around in the air while some will have settled on seats, windows, tables and other things that we're likely to touch.

However, the risk of encountering a cold virus in a given location does not only depend on how many people we're sharing air with. Most people could breathe into our faces all day without doing us a jot of harm because, most of the time, most people are not carrying anything that's going to infect us. The risk of encountering a cold virus in a public place is considerably higher when there is a cold 'going around', to use the colloquial term, because more people are exhaling it. In broad terms, there are more cold viruses in circulation in the winter than in the summer although at the time of writing, the most dangerous of the cold viruses, SARS-CoV-2, has stubbornly refused to adopt a consistent seasonal pattern.

Assessing the risk to ourselves depends very much on the likely consequences of catching a cold. When I make that assessment for myself, I have to consider the stem cell transplant that has left me with a subpar immune system. A cold is likely to leave me debilitated for weeks and I have a significant risk of serious complications if I catch COVID-19. Someone who is not cursed with such a common cold constitution may regard a cold as a minor inconvenience, but the extent of that inconvenience may depend on what they'll be doing in the next week or two. A sniffle that's annoying in a normal week is not something we'll want to bring with us to a significant event like a job interview or a wedding.

Which brings us to the question of what precautions we can reasonably take. Hand hygiene is fairly straightforward; ethanol-based hand sanitisers are available in any pharmacy and they kill off a lot of the viruses we pick up on our hands. Old-fashioned soap and water is more effective against the rhinoviruses although it's not always available because we can't carry a sink in our pockets like a bottle of hand sanitiser. The salient point is that any virus we pick up on our hands can't hurt us until it finds its way off our hands and into our mouth or nose. As long as we clean our hands between passing through a crowded setting and touching our faces, we should be fine – but hand hygiene only protects us against viruses that have already settled on a surface that we might touch.

Facemasks are the only practical way to protect ourselves against viruses floating around in the air. Masking is often more complex than handwashing because we have to balance risk with social acceptability. There is a strong argument for wearing a respirator mask in crowded places during the winter cold season or at any time when there is a high prevalence of COVID-19. Respirator masks are more expensive than surgical masks but not by much. For anyone who risks losing income if they fall ill, a pound or two for a mask that will last the duration of two or three working days may be a sound investment.

There's no such thing as an impenetrable barrier against an airborne virus but hand hygiene and respirator masks are straightforward precautions that we can all take, which should substantially reduce the number of colds we catch every year.

Chapter 14

Air hygiene

April 2020 is a month of bad memories. The COVID-19 pandemic was in the early stages of its global rampage and indoor public spaces were no longer where we went to work, to learn or to socialise because one infected person could flood those spaces with an airborne virus. Governments were responding with restrictions that would have been considered unthinkably draconian only a few weeks earlier.

That was when 36 of the world's leading aerosol scientists tried to warn a World Health Organization advisory panel that they were dealing with an airborne pandemic, and were curtly told they didn't know what they were talking about[*].

All 36 were deeply frustrated by the WHO's attitude. If the WHO's advice was based on the premise that this latest pandemic was not airborne, it would not be good advice. Moreover, they were painfully aware that they were all under lockdown because of a failure of air hygiene.

The rise of Group 36

The term 'air hygiene' arose out of the British Medical Research Council's work during the Second World War, when James Lovelock and Owen Lidwell were tasked with preventing an influenza pandemic emerging in an overcrowded air raid shelter[†]. It's a useful catch-all term

---

[*] Chapter 3 describes the Zoom call in which WHO advisors dismissed the possibility of airborne SARS-CoV-2 transmission.

[†] Chapter 3 describes the formation of the informal Air Hygiene Unit.

for anything and everything to do with preventing the air we breathe from being a conduit for infection.

As a concept, air hygiene has been bouncing up and down the research priority list for the last century and a half. It tends to bounce to the top when a war or a pandemic focuses attention on it but when the crisis has passed, everyone loses interest.

Meanwhile, we've been building an environment around ourselves that is progressively more conducive to viral infection. We've been building that environment for thousands of years, since we learned how to farm our food and enabled our societies to get big enough to sustain cold viruses, but over the last century or so, we've taken it to a whole new level. Most of us – whichever region of the world we happen to live in – now spend a significant amount of time sharing air in the places we go to work, to play, to learn and for many of us, to travel between those places and our homes. Most of those shared indoor spaces have been designed with such scant regard for air hygiene that there's nothing to keep us from sharing airborne viruses along with the shared air.

While most of those viruses have caused nothing worse than colds, most of us have put up with occasional illnesses and soaked up the associated financial costs*. In 2020, governments around the world accepted that we couldn't share public indoor spaces with a virus that hospitalised a quarter of its victims[672] and killed between one in 50 and one in 200 of them[673].

In so many of our buildings, the air hygiene was so bad that the emergence of one unusually dangerous virus was all it took to make each building a potentially lethal environment to enter – and it had long been known that such a virus would inevitably emerge.

After that Zoom call with the WHO, Lidia Morawska decided that enough was enough.

Morawska started her career as a nuclear physicist in her native Poland. She later moved to Toronto to study radon, a radioactive gas that can be a natural pollutant. She found that when radon causes harm, it's because it attaches to airborne particles fine enough to be inhaled down to the lungs. It followed that the concentration of such particles

---

* Several hundred pounds, dollars or euros every year according to my back-of-envelope calculations in Chapter 5.

in the air was an important factor in the harm caused by radon, so she looked into the published literature on the subject – and didn't find it. At the time, nobody had looked into what particles were floating around in the air we breathe.

'I said, wow, what is this?' she told me. 'There were no papers, no literature on this. And I thought, well, this is an important and fascinating science.'[674]

Her next move was to Brisbane's Queensland University of Technology, where she made such 'ultrafine particles' the main focus of her research. It was an unglamorous subject which is probably why the WHO advisor who talked over her appears to have had no idea he'd just interrupted a leading expert on the subject.

It was the sort of ignorance and intransigence that's deeply frustrating to come up against but sometimes, frustration inspires cohesion.

For Morawska in Brisbane, that conversation happened on a Saturday evening. On Sunday, Morawska turned her frustration into action which, for a scientist in lockdown, meant that she wrote a letter. She contacted the other aerosol scientists who had been on that call and found every one of them as keen as she was to take the matter further. Three days later, all 36 aerosol scientists from 14 different countries had agreed on the text of the letter, signed it and sent it to the Director-General of the WHO[675].

It was the opening shot of 'Group 36', as the aerosol scientists called themselves. They didn't want their lobbying to turn into a public slanging match, so they kept their media engagement to a minimum. Nevertheless, air hygiene became so important that *TIME* magazine named Morawska as one of the TIME100 most influential people of 2021[676]. More important to Morawska was that Group 36 eventually persuaded the WHO to change its position and national health authorities started to follow its lead*.

At the same time, Group 36 came up with a plan for how we might be able to do something about our buildings being so well designed for airborne virus transmission[677]. They came up with a three-part hierarchical framework that would allow us to learn, work and be

---

* Chapter 3 describes the WHO's belated change of position in more detail.

entertained without picking up the colds – or worse – that we take home to infect our nearest and dearest.

The first and best option is to remove viruses from the air before anyone has a chance to inhale them.

Unfortunately, many modern buildings have been designed with such poor ventilation that it won't always be possible to get rid of virus-laden air fast enough. If the people in a room are going to have to breathe each other's exhalations, Group 36 suggested the second-best option of disinfecting the air.

Only when neither option is achievable is it worth considering the third option: giving up on sharing our indoor spaces and going into 'lockdown'. Unfortunately, air hygiene has been such a low priority for so long that in most of the world's buildings, neither of the first two options could be implemented fast enough to deal with a fast-spreading pandemic. In 2020, lockdown was the only option.

Viewed from the midst of that third option – which, in a better planned society, would have been the last resort instead of the only thing to do – it was particularly galling that the first option had been recognised a century and a half earlier, by the first person to apply mathematics to human health.

The first air hygienist

To a Victorian Briton, a hospital was not the place of healing that we think of today but a last resort and in many cases, a last stop for those who had nowhere else to die.

In 1863, one frustrated Victorian wrote, 'It may seem a strange principle to enunciate as the very first requirement in a Hospital that it should do the sick no harm. It is quite necessary, nevertheless, to lay down such a principle, because the actual mortality *in* hospitals, especially in those of large crowded cities, is very much higher than any calculation founded on the mortality of the same class of diseases among patients treated *out of* hospital will lead us to expect'[678].

It's a remarkable passage. The reference to the calculation reveals that the conclusion is underpinned by some of the first medical statistics, making it a seminal passage in the embryonic disciplines of public health and epidemiology as well as air hygiene. Yet it's no abstract exercise in description. It's the beginning of a treatise on turning hospitals into places of healing rather than places of misery and death.

But then, Florence Nightingale was a remarkable person.

She first came to prominence in 1854 when she recruited and led a team of nurses to work in the British army's hospital in Scutari*, which was inundated with casualties from the Crimean War battlefields. Nightingale's care for those wounded soldiers made her a national heroine who was celebrated as 'the lady with the lamp' for her willingness to do whatever needed doing at all hours of the day and the night.

Despite her fame, there was one person who believed her work in Scutari did not deserve celebration: Nightingale herself. She believed the value of her work in Scutari lay in neither the effort that she put into it nor the headlines it garnered but in what it achieved. Being one of the most accomplished mathematicians of her day, she was able to precisely calculate those results.

She wasn't impressed.

Disease had been so rife in the Scutari hospital that a soldier sent there for treatment was far more likely to die than if he was left to the tender mercies of the regimental surgeons treating soldiers on the frontline. One such disease, probably a bacterial infection called brucellosis, ended Nightingale's tenure in Scutari and sent her back to Britain. Nightingale never fully recovered but despite her fatigue and intermittent malaise, she committed the rest of her life to sanitary reform[679].

Nightingale wasn't the only Victorian advocating for sanitary reform, but she was the only one who backed up her arguments with rigorous mathematics. Her application of statistics to human affairs marked the beginning of the modern discipline of public health and posthumously earned the 'lady with the lamp' the nerdier nickname of the patron saint of statistics. We can only guess how Nightingale herself, a staunch protestant, would regard her tongue-in-cheek canonisation.

She took the first step toward that nickname when she wrote *Notes on Hospitals*, a treatise in which she argued that hospitals were designed 'to lead to difficulty in ventilation, or want of light'[680], making them hubs of the sort of pestilence they were supposed to combat.

---

* Now Üsküdar, Turkey.

Her solution was to rethink how hospitals were designed. Instead of a single large building, Nightingale proposed that a hospital should be composed of several narrow 'pavilions' that enabled every bed to be placed close to a window which, before there was electricity to power lights and fans, was the main source of both ventilation and natural light. She stated that air should pass through the pavilions at a rate of 1,500 cubic feet per hour per patient, once again showing her flair for mathematics by giving a precise figure where her contemporaries tended to give vague descriptions.

Five years after *Notes on Hospitals* was published, Queen Victoria herself laid the first stone of the first hospital to be built to Nightingale's specifications. St Thomas's Hospital, across the River Thames from London's Houses of Parliament, incorporated another of Nightingale's innovations: a nursing college, which began the process of professionalising nursing.

St Thomas's remains one of London's largest hospitals, although Nightingale's pavilions are no longer a part of it. They were heavily damaged by bombing during the Second World War and in the 1960s, they were demolished and replaced by a brutalist concrete block. Nightingale's legacy remains in a museum dedicated to her but less explicitly, if perhaps more pervasively, in the electric fans that ventilate the building and in the highly qualified nurses who run its wards – and those of the overwhelming majority of hospitals and indeed all large buildings in the world today.

From Scutari to SARS-CoV-2

If Nightingale's ideas carried enough weight to govern the design of hospitals in the 1860s, it's perplexing that a WHO panel was so strongly opposed to essentially the same ideas in 2020. Alexander Langmuir's vigorous opposition to the concept of airborne infection is part of the reason*. Another factor is that Nightingale advocated ventilation for its own sake rather than as a way to prevent airborne infection. She saw no distinction between those two goals because at the time she wrote

---

* Chapter 3 covers Langmuir's objections to airborne infection, which became so much more influential than his eventual refutation of them.

Notes on Hospitals, she subscribed to the miasma theory that noxious air is the direct cause of disease. She dismissed the possibility that patients could be infecting each other, arguing that 'there is no proof, such as would be admitted in any scientific inquiry, that there is any such thing as "contagion"'[681].

In 1863, she wasn't wrong; the microbial theory of disease still lay a decade in the future. Moreover, stagnant and stinking air was and is unhealthy, whether or not one understands how it's unhealthy.

Nightingale's ideas were first applied to British hospitals, but they were adopted in many other countries. By the late nineteenth century, they had moved into the domestic sphere when German architects designing homes adopted the guiding principle of *Licht und Luft*, meaning 'light and air'. By the 1920s, modernist architects like Charles-Édouard Jeanneret, better known by his *nom de guerre* of Le Corbusier, and Walter Gropius, founder of the German Bauhaus school, were designing homes with large windows and balconies[682] that admitted all the light and air that Nightingale might have wished. With tuberculosis riddling European cities and the influenza pandemics of 1889 and 1918 far from forgotten, if rarely talked about, such homes became very popular – at least among those who could afford them.

As buildings became larger and more complex, ventilation became a progressively more important design consideration, but it was not driven by concerns around airborne infection. Even today, ventilation systems tend to be designed more to prevent the build-up of nasty smells than to prevent airborne infection.

While *Licht und Luft* guided the design of the indoor domestic space, it rarely penetrated the design of indoor public spaces like schools and workplaces which, as Owen Lidwell would show in the 1950s*, became hubs of viral exchange. Attempts to address the problem were sporadic and usually driven by concerns that proved ephemeral.

The MRC's Air Hygiene Unit is a typical case in point. Concerns about overcrowding generating an influenza pandemic were misplaced but there were good reasons to be concerned about air hygiene in air raid shelters. However, it's revealing that the air hygiene of the London

---

* Chapter 5 describes Lidwell's groundbreaking epidemiology in rural Bowerchalke and later in the urban context of London and Newcastle.

Underground network only became a concern during the Second World War. The 'tube' has always been notorious for overcrowding and poor-quality air and, as Glenn Miller's big band jazz and Lonnie Donegan's skiffle gave way to the rock 'n' roll of The Beatles and the Rolling Stones, air on the tube remained as foetid as ever. Christopher Andrewes once observed that it was just as well that colds are seasonal because otherwise, 'London Transport would ensure an all-round-the-year cold epidemic'[683].

The Air Hygiene Unit's interest was driven less by concern for the health of the individuals taking shelter than by the exigencies of war. With much of the able-bodied population in uniform, it mattered that dockers, machinists and service personnel on leave remained healthy enough to go back to work once the all-clear was sounded. When peace returned, it became much easier to replace anyone who became too ill to work and air hygiene on the tube ceased to be a priority.

The influenza pandemics of 1957 and 1968 spurred another burst of enthusiasm for air hygiene right in the middle of the cell culture-driven boom in cold virus discovery. For a while, the study of air hygiene became almost synonymous with the discovery of cold virus transmission. Colds may not be a major health problem in themselves, the reasoning went, but they offered a way to study the pathways of pandemic influenza transmission when there was no pandemic to study.

By 1986, as the enthusiasm driven by those pandemics was waning, Anthony Arundel and his colleagues at Vancouver's Simon Fraser University set out to distil some principles of airborne transmission out of the reams of research papers that had been published by then. It's a paradox of scientific research that Arundel needed a huge volume of research to establish a few succinct principles. However, such principles are defined by holding true across multiple contexts so until research had been done in multiple contexts, there was no way to tell the difference between a general principle and something that only appeared in one of those contexts. Arundel needed reports from many different schools, military bases, workplaces, universities and wherever else scientists had followed colds around to see which factors were universal.

He came up with six factors[684] that govern the transmission of colds, or indeed any other infection carried by airborne respiratory droplets, which are worth stating in full:

1. The number of infected people producing contaminated aerosols.
2. The number of susceptible individuals.
3. The length of time for which individuals are exposed.
4. The ventilation rate in the local indoor environment.
5. The settling rate of contaminated aerosols.
6. The survival of pathogens attached to aerosols.

If the late 1980s had seen another world war or a pandemic to focus attention on airborne transmission, perhaps those principles would have underpinned a drive to do something about them. In the event, air hygiene once again fell off the agenda and the world went on building cities that could have been designed to the specifications of an ambitious rhinovirus.

Carbon dioxide in classrooms

In 2013, the British Secretary of State for Education, Michael Gove, suggested removing Florence Nightingale from the national curriculum[685]. More than a century after her death in 1910, the lady with the lamp still commanded so much affection that Gove's idea provoked irate newspaper columns, even more irate social media posts and a petition to Parliament[686].

Perusing the arguments over Nightingale's legacy, it quickly becomes apparent that the most treasured aspect of her legacy relates to her year in Scutari rather than her subsequent decades of advocating for sanitary reform and ventilation. In the minds of most of her champions, she remains the lady with the lamp rather than the patron saint of statistics. Her advocacy for air hygiene had slipped from the collective memory.

By the beginning of the twenty-first century, air hygiene had been subsumed into disciplines like building physics and aerosol science. Much of the focus had shifted from hospitals to schools where, at least in Britain, there was far more willingness to learn *about* Nightingale than to learn *from* her. Many British children first learned the name of Florence Nightingale in rooms so badly ventilated that Nightingale would have taken one sniff and ordered the windows opened immediately – although, by then, she would not have needed to rely on her nose to assess the air in a room.

A twenty-first-century air hygienist uses sensors that give second-by-second recordings of a range of parameters including the relative humidity that governs airborne virus transmission, the so-called PM2.5 particles that are small enough to pass through the windpipe to irritate the lungs and, of most interest from the perspective of ventilation, the concentration of carbon dioxide.

Carbon dioxide matters because our metabolism converts oxygen into carbon dioxide. When the air in a room is passing through the lungs of the people in that room faster than it's passing through the room, the carbon dioxide level will rise as those people turn the room's air into carbon dioxide. It's easier to measure carbon dioxide than oxygen, so the former tends to be used as a measure of how much of the air in a room has been through someone's lungs.

Carbon dioxide is measured in parts per million of the air. Outdoor air contains around 400 parts per million but let two dozen children into a poorly ventilated classroom – which is also a typical classroom – and watch the reading from a carbon dioxide monitor, you'll see how fast two dozen pairs of little lungs drive up the carbon dioxide levels and, one can infer, drive down the oxygen level. It won't take long for the carbon dioxide to hit 1,000 parts per million, at which point the children will start to lose focus. Their cognitive impairment will steadily worsen as the carbon dioxide continues to rise[687]. If they're doing a mathematical exercise, they'll make more mistakes with their arithmetic. If they're writing essays, they'll be putting words on the page more slowly. If they're trying to follow what their probably drowsy teacher is saying to them, they'll be more easily distracted from it.

Among the various studies of the world's classrooms done in the late twentieth and early twenty-first centuries, carbon dioxide levels below 1,000 parts per million are the exception rather than the norm. To give a couple of examples, one assessment of schools in Aberdeen found that carbon dioxide levels as high as 4,000 parts per million were not unusual[688] while a study of schools in the south-western USA frequently measured carbon dioxide levels that topped out their monitors at 6,000 parts per million, meaning that the levels in the classrooms were likely to be even higher[689].

By 2020, there were enough studies from different schools in different countries that Pawel Wargocki, an indoor climate scientist at the Technical University of Denmark and a member of Group 36, was able to carry out a meta-analysis. He and his colleagues found that they could see the effect of all that carbon dioxide in schools' test results. They calculated that dropping carbon dioxide from the all-too-commonly measured 2,400 parts per million to the depressingly rarely measured 900 parts per million would enable children to score around five per cent higher in whichever national tests they had to take[690].

Wargocki also found that high levels of carbon dioxide in classrooms were associated with children not coming to school at all. For every increase of 100 parts per million, absences increased by one day every two-and-a-half years for every child in the school. That means that at the oft-recorded 1,500 parts per million, which is 1,100 above outdoor air, that's two or three days per year per child. At 3,000 parts per million, well below the highest levels recorded in both the Aberdeen and southwestern American studies, that's more than a dozen days off every year attributable purely to poor ventilation.

Those absences aren't likely to be caused by the heightened carbon dioxide or even the low oxygen levels that go with it. Carbon dioxide build-up makes children dozy and distractible but a lungful of outdoor air is all it takes to fix that. The carbon dioxide level is an indicator of how much of a room's air has been breathed in and out of someone's lungs and with every exhalation of carbon dioxide comes an exhalation of fine droplets. If one of the children is lacing those droplets with cold viruses, the high carbon dioxide indicates that those virus-laden droplets are trapped in the room with the children.

When Wargocki looked for an association between absence and carbon dioxide, he was looking for an easily measurable indication of how many respiratory droplets were floating around in that room.

When the COVID-19 pandemic prompted another burst of interest and, more importantly, funding for air hygiene, Group 36 were able to draw on studies like Wargocki's to make some specific carbon dioxide-related recommendations. As long as the carbon dioxide was no more than 800 parts per million, which is twice the typical outdoor level, Group 36 calculated that anyone inside that room should be reasonably safe from even the highly infectious COVID-19 virus[691]. They also recommended a ventilation rate of 14 litres per second for every person in the room, which was only marginally higher than Florence

Nightingale had recommended back in 1863* without the aid of any sensor more sophisticated than her nose.

The Class-ACT study

For most schoolchildren, classroom air containing only 800 parts per million of carbon dioxide is a distant aspiration. In mid-latitude countries like Britain, the situation is exacerbated by the highest carbon dioxide levels – and concurrently the highest airborne infection risk – being during the winter cold season. That's because most classrooms are ventilated using the old-fashioned way: by opening a window.

In many classrooms, or indeed in many rooms in most buildings, that's all it takes to drop the carbon dioxide below the 800 parts per million threshold and flush out virus-laden droplets. However, most classroom windows remain closed throughout the winter to retain whatever heat is generated by the usually clapped-out heating systems. It's still worth occasionally opening a window for a few minutes. In a brick or concrete building, most of the heat energy is in the building's structure, not the air, so it's possible to let some outdoor air in without noticeably adding to the heating bill.

However, building managers have been known to lose their sense of humour when they catch anyone doing anything at all likely to raise the energy bills. One air hygienist told me that when he looks at the windows of British classrooms, he often finds so much paint slapped over them that they're permanently gummed shut[692].

The upshot is that even when windows aren't painted closed, they will almost always *be* closed during the winter: the season when someone is most likely to be exhaling cold viruses into the classroom and the heating drives down the relative humidity to keep them airborne†.

If someone read Arundel's six principles as an instruction manual for infecting as many people as possible, they would design something that looks like a typical British classroom.

---

* Nightingale's recommended ventilation rate of 1,500 cubic feet per person per hour is equivalent to 11.8 litres per person per second.

† Chapter 5 describes how heating enhances airborne virus transmission by lowering the relative humidity.

To apply Group 36's framework, the first question would be whether it might be possible to improve classroom ventilation without turning children blue with cold or driving school budgets into the red. For most classrooms, the answer is going to be no. Fitting a network of fans and ducts in a building that was never designed for them is always expensive and not always practical.

If Group 36's first option of ventilation isn't possible, their second-best option is disinfection. If we really can't give schoolchildren enough oxygen, perhaps we can at least limit how many viruses they inhale with all that carbon dioxide. In 2021, two groups of researchers put their heads together to explore the possibility. One group was led by Catherine Noakes, an engineer at the University of Leeds and a member of Group 36. The other was led by Mark Mon-Williams, an educational psychologist whose time was split between the University of Leeds and the Bradford Institute of Health Research.

There were a lot of theoretical modelling studies showing that it *should* be possible to disinfect classroom air but, as Noakes and Mon-Williams were painfully aware, it's notoriously difficult to code the chaos and unpredictability of a classroom seething with small children into anything logical enough to be run on a computer. Until someone ran a field study, it was impossible to know whether disinfection *did* work in the real world as well as in a theoretical model.

The lack of real-world data on air disinfection was due to that perennial problem of air hygiene research: it takes a crisis to persuade anyone to fork out the funding needed to do a field trial. In 2021, with COVID-19 causing such a crisis, Noakes and Mon-Williams persuaded the government to loosen its purse strings and fund a much-needed field trial. The classroom air cleaning technology study, or Class-ACT, was born.

For once, cold researchers had come up with a name that was both memorable and pronounceable.

Mon-Williams was one of the leaders of the 'Born in Bradford' programme of research into many different aspects of family health, including those that related to schools. He had been working with Bradford schools for several years before COVID-19 arrived, so he already knew which headteachers to ask if they'd be willing to join a trial of air disinfection.

The original plan for Class-ACT was to try out two different approaches to disinfection. Some schools would be fitted with so-called high-efficiency particulate air filters, or HEPA filters, which use a fan to draw air through a mesh of filters. The principle is similar to that used in facemasks, although HEPA units have far more layers of filters than would be practical for a facemask.

Other schools would be fitted with upper-room ultraviolet, which was essentially the same set-up that William and Mildred Wells's Germantown experiments used to prove airborne virus transmission in the 1930s*. The Class-ACT team fitted ultraviolet bulbs above baffles that protected students and teachers from ultraviolet rays, depending on children's body heat to carry airborne viruses above the baffles to be zapped.

A third group of schools were fitted with neither HEPA filters nor ultraviolet and were monitored to see what happened in the absence of either.

It looked like an ideal set-up for a field trial until Class-ACT stumbled into a bureaucratic quagmire. They had installed the upper-room ultraviolet according to the relevant workplace regulations but in Britain, schools fall into a different regulatory category. Amid the reams of paperwork defining what can and cannot happen in a classroom, there was no guidance on ultraviolet lamps.

In the absence of any rules, school administrators baulked at switching on the virus-murdering ultraviolet bulbs. The Class-ACT team became embroiled in several months of back-and-forth over whether systems that were perfectly safe and legal in an office or workshop posed any risk in a classroom.

Ironically, the regulatory gap preventing Class-ACT from disinfecting classroom air was a big part of the reason why the air needed disinfecting in the first place. The same workplace regulations mandate minimum ventilation standards and it's because they don't apply to schools that British classroom air is of such poor quality. Just as nobody had considered that schoolchildren might need as much oxygen as working adults, nobody had thought to draw up any guidelines for a system that was fully approved in adult workplaces.

---

* Chapter 3 describes the seminal Germantown experiments.

Eventually, everybody was satisfied that ultraviolet rays that could not reach children could not harm them but by the time Class-ACT received the go-ahead to switch on the bulbs, there was so little of the school year left that it was too late to extract any meaningful results[693]. Consequently, the Wellses' experiments of the 1930s remain the most recent trial of ultraviolet disinfection in schools.

The trial of disinfection by filtration went better. Class-ACT's HEPA filters whirred away throughout the 2021–2022 school year. The multilayered filters would capture droplets containing any virus but at that time, the most pressing question was whether disinfection could prevent COVID-19. In the first two months of 2022, that question was answered when the SARS-CoV-2 Omicron variant arrived in Britain. It was considerably more infectious than the earlier variants and it swept through schools like the proverbial dose of salts.

Yet another wave of COVID-19 gave Class-ACT a lot of data. Among children breathing HEPA-filtered air, absences were between three-quarters and four-fifths of the level among children whose air was not disinfected.

The Class-ACT team then looked into their secondary objective, which was to see if HEPA filters protected children against other cold viruses. They looked at the absences before and after the Omicron wave and found that, predictably enough, there were fewer absences in all classrooms than when Omicron was on the rampage, but children breathing HEPA-filtered air still missed less school than children breathing unfiltered air[694].

The Class-ACT study had shown that filters help to keep children healthy and at school – up to a point. Most of the schooldays that would have been lost without the filters were still lost with them, although even the most fanatical air hygienist wouldn't expect them to have prevented every absence. Even if those filters could snag every virus in the classroom before it got breathed up a child's nose – and it was never realistic to expect them to work that well – they could only protect children while they were in class. Many absences would be caused by colds caught outside the school or by illnesses that had nothing to do with airborne viruses.

When considering absences caused by colds, we need to consider that some children will account for a lot more of those absences than others. Just as some adults have more of a common cold constitution, so cold viruses will affect some children worse than others. Asthma affects more than one in every 20 British children[695], meaning at least one in a typical class. While the other 19 children might not be slowed down by having a sniffle, colds are likely to aggravate that twentieth child's asthma so he'll be absent a lot more than the others.

Asthma is one of many factors that predispose a child to a cold nasty enough to keep them at home and many of those factors disproportionately affect children from less wealthy families. Socio-economic deprivation, to use the technical term, often means living in an area with a lot of air pollution and living in homes infested with black mould releasing spores into the air[696], both of which exacerbate colds and asthma alike. Moreover, families that are struggling to put food on the table can't be too picky about what that food is, making their children more likely to be obese, which in turn makes them more prone to the sort of inflammatory response that defines a bad cold[697]. Sanchita Pal, a community paediatrician at Evelina London Children's Hospital, right next to St Thomas's, described the combination of factors to me as a 'nice mixing pot'[698] making less wealthy children more vulnerable to colds and more likely to be absent often enough to fall behind at school.

That's the situation of many of the children who attended the Bradford schools where the Class-ACT study took place. Mark Mon-Williams told me that many live in what social scientists call 'low-income multigenerational households', meaning that those children share their home with aunts, uncles and grandparents as well as their parents, and the entire household is struggling to make ends meet. In such a household, every cold that a child brings home from a poorly ventilated classroom is another problem for a household that already has problems to spare.

The working-age adults are probably in jobs that pay by the hour rather than by the month. If they're too ill to work, they don't get paid, which can be the difference between paying the electricity bill and being cut off. Moreover, a bout of influenza or COVID-19 may not be a trivial matter for ageing grandparents whose health is already failing.

'If people start perceiving schools as unhealthy, unsafe environments, they're less inclined to send their children to school,' was how Mon-Williams explained it to me. 'We've currently got a school absence crisis playing across our district, 30,000 children persistently absent from school.'[699]

This is where decades of neglecting air hygiene across our indoor environment in general, and in schools in particular, had brought us. Parents are beginning to see schools less as places of safety and learning than as the places where children bring home viruses that keep mum and dad from their jobs and send grandma and grandad to hospital.

Class-ACT showed that HEPA filters can improve the situation, although they revealed a few problems of their own. Disinfecting air that children are constantly breathing viruses into requires all the air in the classroom to pass through a HEPA filter between three and six times every hour which, in a typical classroom, required three units. Most teachers will tell anyone who cares to listen that their classroom does not have enough sockets, making them less than enthusiastic about committing three of them to HEPA filters. Class-ACT also found that HEPA units are not designed to survive one of the most destructive forces on earth: a class of inquisitive eight-year-olds.

'Kids like to fiddle with the HEPA filters,' Christopher Brown, Class-ACT's project manager, told me. 'Quite a lot of them, you turn them on, they've got Lego bricks flying around inside them.'[700]

Perhaps the biggest problem was that the HEPA filters made enough noise to interfere with teaching. Class-ACT had no way to record how often teachers turned the HEPA fans to a lower setting or turned them off altogether, but it's unlikely that they were always dialled up to the requisite three to six filtrations per hour. On the other hand, the 'turbo' mode that generated the most noise proved popular in classrooms that became uncomfortably hot during the summer, where the HEPA filters ended up being used as fans.

Despite those problems, the Class-ACT study found that many schools, if not quite all of them, did not want their pupils breathing infected air. For every school that asked for its filters to be removed, there was always another school that was more than happy to take them on.

Air hygiene today and tomorrow

As I write this, Class-ACT remains the only evaluation of how air disinfection prevents airborne infections that involved multiple buildings. However, it leaves many questions unanswered. Most obviously, could the upper room ultraviolet have improved on the benefits of the HEPA filters without taking up those extra sockets, making a distracting noise or being vulnerable to Lego-slinging children? We don't know the answer because they were never switched on.

Class-ACT did not evaluate 'active' ultraviolet units, in which fans draw air through an intense blast of ultraviolet. The drawback is that the active ultraviolet units currently on the market are even noisier than HEPA filters, which is why Noakes and Mon-Williams didn't think they would be appropriate for classrooms. They may have a place in some workplace settings or on public transport where there's already a certain amount of background noise but at the time of writing, there has been no field trial to establish how they compare to HEPA filters.

The elephant in any room in which air needs disinfecting is the cost of disinfection. Class-ACT's HEPA filters cost around £500* [702] plus the electricity needed to run them. They also added to the already substantial workload of the schools' maintenance staff, who needed to clean the filters every half term.

Class-ACT's funding covered those costs while the trial was ongoing but outside the context of a trial, air disinfection may be beyond what a typical British school's budget can absorb.

Nor should we forget that, as Morawska, Noakes, Wargocki and the other air hygienists whom the World Health Organization advisors dismissed, disinfection is only the second-best option[703]. The best is ventilation, which incurs costs of its own but offers even more benefits.

Taking all the issues together leads to a simple question: would a reasonable society leave the quality of its children's air to already-overstretched school budgets or would it baulk at the idea of herding its children into poorly ventilated infection chambers?

My view is that it would not.

If keeping children healthy enough to attend school in classrooms sufficiently oxygenated for them to learn something isn't a good enough reason to do something about the air quality, then we can also consider the economic costs inherent to absence from work and visits to GPs and hospitals.

---

* In 2025, that's $686 or €583[701].

Framing air hygiene as an investment rather than an overhead is the sort of thing that makes economists frown and demand cost-benefit analyses. At the time of writing, the only such analysis comes from a collaboration between air hygienists at the University of Brescia and the Polytechnic University of Milan, who tested filters in several schools in Milan during the 2023–2024 school year. Their focus was on air pollution rather than infection[704], although the filters they used would have captured infective droplets at least as efficiently as Class-ACT's HEPA filters. The Milan team calculated that preventing one child missing one day of school worked out at only €10.60* [706].

Factors affecting absences will vary from one part of the world to another, but the cost is likely to be in the same ballpark as in Milan – and €10.60 per day doesn't sound like an unreasonable ballpark. It sounds like a bearable investment in the future of the children affected by the school absence crisis that worried Mon-Williams[707].

Despite decades of air hygiene being neglected, the air hygienists I spoke to were a lot more optimistic than I'd anticipated. Lidia Morawska pointed out that Belgium is introducing regulations that will mandate maximum carbon dioxide levels in indoor public spaces[708], which effectively mandates a decent level of ventilation. Where Belgium leads, she hoped that others will follow.

As Mark Mon-Williams put it, 'The arc of history suggests that actually, good quality … science … takes a long time. But eventually it starts to … shape things … our job is … to … keep saying, well, here's the evidence right?'[709]

Florence Nightingale would have approved.

---

* In 2025, when the working paper was published, €10.60 was equivalent to £9.10 or $12.50[705].

Epilogue

We can do better than the Two-hat Remedy

Scientists who work on a given topic are engaged in an ongoing conversation with one another, sharing their results and their opinions as they work out where to take their research next. When individuals come together over a shared interest, they form a community and when a community has lasted for as long as the community of cold researchers – a century and counting since Alphonse Dochez's seminal experiments proving colds to be caused by viral infection – that community develops a folklore.

Listen to scientists talking over beers in a conference centre bar and sooner or later, one of them will mention something they've heard somewhere from someone who heard it from someone else. They'll smile as they say it, inviting you to laugh, but you can tell they're wondering if there *might* be some truth to it. Those anecdotes won't appear in research papers, where every statement needs to be backed up by evidence, but the cold research community's myths sometimes turn up in memoirs or public lectures.

One such myth is the two-hat remedy, which is attributed to a different source every time it appears. Nobody who wrote it down claimed to believe in it, but they all insist that somebody, somewhere, believed that if you are afflicted by a cold, you should hang your hat on your bedpost, go to bed and hit the bottle until you see two hats. Next morning, you'll wake up and find your cold is cured.

Christopher Andrewes, the co-discoverer of the influenza virus and founder of the Common Cold Unit, wrote that he heard about the two-hat cure on a visit to Hungary[710]. His successor as CCU director, David Tyrrell, claimed to have read it[711] in *Domestic Medicine*, a late eighteenth-century guide to home-made remedies for every imaginable malady written by a Scottish physician called William Buchan.

Domestic Medicine was the blockbuster bestseller of its time. No Georgian British bookshelf was complete without a copy and, through translation, *Domestic Medicine* became a fixture in households across Europe. It was so widely celebrated that Buchan received a letter of thanks and a gold medal from the empress of Russia[712].

However, despite the insistence of the usually meticulous Tyrrell, *Domestic Medicine* does not describe the two-hat remedy* but rather advises *against* treating colds with 'strong liquors'[713]. Buchan's favoured cold remedy was to place one's legs in warm water to induce sweating and to raise blisters on the back, which would be as ineffective as the two-hat remedy and sounds a lot less fun.

Another version appears in *The Plague of the Spanish Lady*, Richard Collier's collection of reminiscences of the 1918 influenza pandemic. One of Collier's correspondents wrote of an incident in the French city of Lyon when a medical orderly was called to the home of an influenza-stricken weaver. Lacking any effective treatment, the orderly told the weaver to drink sugared red wine until he saw not two but three hats 'and *voilà!*, you'll be cured'[714]. The reminiscence ended there, without establishing whether the weaver recovered.

The two-hat remedy is an embellishment of the age-old belief that the best treatment for a cold is a stiff drink or, perhaps, of having a cold being an age-old excuse to get drunk. Presumably that's why Buchan felt the need to include specific advice against boozing away colds in *Domestic Medicine*.

Two and a half centuries of research later, belief in the medicinal properties of alcohol is alive and well. When I was a young researcher, I once left my laboratory early because I had a cold and bumped into one of the world's leading experts on viral diseases. Summoning his decades of experience, he sternly advised me to go to bed with a stiff whiskey – although he didn't mention hats.

If the story of the French orderly in 1918 is to be trusted, the two-hat myth predates Dochez's foundational research that showed colds are

---

* I must add the caveat that I haven't been able to check all 19 editions of *Domestic Medicine* or to establish which of them Tyrrell was referring to. It's not impossible that Buchan had an epiphany for or against the medicinal properties of heavy drinking in between editions although it doesn't seem likely. Sweating, blistering and abstention were commonplace treatments at the time.

caused by viruses. A century on from Dochez's work, we have better treatment and prevention options than trying to drink a cold away, but we still don't have a reliable cure. More importantly, we're still not very good at dealing with the more serious conditions that cold viruses cause. The introductory chapter related the tribulations of Lucy the fitness instructor, who has been left seriously impaired by seasonal influenza and later SARS-CoV-2, both of which are widely regarded as cold viruses that we must live with.

Like most people struggling with long-term debilitation caused by cold viruses, Lucy finds such insouciance frustrating.

'The scary thing is that I am seriously not alone,' she told me. 'You know, I'm on a lot of Facebook groups and a lot of Reddit groups and there are a lot of people out there that are getting the symptoms after Covid that I got after the flu.'[715]

Her fear is supported by solid data. As I write this, the most recent estimates for both Britain[716] and the USA[717] were that more than one person in every 40 has long covid.

Viruses that ride floating droplets around our buildings remain an intractable problem. Moreover, the number of colds we catch is a direct measure of how vulnerable we are to the next airborne pandemic virus that emerges. We can't know when that will happen. It may be 50 years from now or it may be tomorrow but we know that it's going to happen and the more complacent we are about colds, the harder the next pandemic will hit us.

It doesn't have to be this way. Most cold viruses can only exist because we've spent the last few thousand years building a society that enables their existence. It follows that with a little intention we can make our society a lot less friendly for them. Air hygienists from Florence Nightingale to Lidia Morawska have shown us how to remove a lot, if not all, of the viruses infesting the air we breathe. If we choose to learn from them, we'll lose fewer days to illness, our children will not be learning their lessons in the infection chambers that too many schools have become and we'll be a lot less likely to have our lives changed as dramatically as Lucy's.

We know what we need to do. The choice facing us now is whether we do it.

Sneeze

The experts' remedies

For the time being, the occasional cold is unavoidable, so I asked some of the world's leading experts on the subject what they do when they catch a cold. Here's what they told me:

Ronald Eccles, professor emeritus at the School of Biosciences and former director of the Common Cold Centre, Cardiff University, Cardiff, UK:

'Once I know I'm getting a cold, The first thing is hot, tasty drinks. They're beneficial … they're harmless and very good for sore throat and cough, very good … you can take them as often as you want. If you're taking a standard cough medicine, it's every four to six hours, whereas with a hot drink you can get up in the middle of the night. Take one, get up again, give it to the children, etc. It's harmless. and it's really the old traditional honey and lemon. That's the tradition … My second treatment would be an analgesic: paracetamol, ibuprofen or aspirin. I usually do take paracetamol because that will control the pain symptoms, and any muscle aches and pains, feverishness, chilliness, sore throat, pain, headache, etc. So you've hit a lot of symptoms with that. And then, finally, I would take a topical, nasal, decongestant: xylometazoline, oxymetazoline … to open up the nose because you cannot sleep with a blocked nose. You're going into periods of sleep apnoea, and you wake up feeling awful in the morning. That's because you've had a terrible night's sleep. So that's my recommendation. A triple approach to the common cold.'

Ron Fouchier, professor in molecular virology and deputy head of the Department of Viroscience, Erasmus Medical Centre, Rotterdam, Netherlands:

'I stay home … because I don't want to spread it through … the people in my lab … and then of course, you have to watch your hygiene with how you cough and sneeze. If I have fever, I take paracetamol … to try to repress my fever. But that's it.'

Stephen Griffin, professor of cancer virology at University of Leeds School of Medicine, UK, and co-chair of Independent SAGE:

'The first thing I'll do is test for SARS[-CoV-2] ... I'll test multiple times because ... you need to keep testing over a sustained period. In terms of avoiding infection. I'll wear a respirator in crowded indoor venues. I'll wear a respirator in public transport ... and I keep an eye on the epidemiology ... I do my best to avoid ... direct exposure, but I balance it out with the fact that I know that I've made a very robust vaccine response to SARS[-CoV-2]. I get my flu jab every year. I'm old enough that I've had most rhinoviruses so I don't mask anywhere near as much as I used to. I used to mask at work religiously. I don't do that any more, but that's partly because our building's not as busy as it used to be ... it's about ... balancing the risk with what it is you want to be doing. You know, if you want to go to a restaurant and it's high prevalence, think twice ... and if I'm ill, I don't go places. I don't mix ... If I know it's just a standard respiratory virus, I might do a little bit more than if it was SARS[-CoV-2], for example, but I wouldn't do anything if it was covid positive ... When I'm not well ... Night Nurse* is very good for helping ... the best management you can have for those sorts of things is to keep hydrated, take painkillers when you need them and try and rest. People underestimate these things and ... if you rush back into it – and especially SARS[-CoV-2], actually, if you try and come back too quickly from covid, it can really do you in ... it definitely takes a lot out of you, even if you're vaccinated.'

Ruth Harvey, assistant director of the Worldwide Influenza Centre, Francis Crick Institute, London, UK:

'Oh, goodness! What do I do to prevent them? I try to make sure that I wash my hands as soon as I get off public transport because washing your hands is a very good thing to do. I have my vaccines regularly, of course ... If I get one, I stay at home and feel very sorry for myself under a blanket with my cat and probably make my husband's life a misery by grizzling a lot ... I've always been very much a fan of keep

---

* Brand name for a mixture of paracetamol, promethazine and dextromethorphan.

your cold to yourself. I get ... frustrated by Lemsip adverts where it's like, 'have a Lemsip and get on with your day'... Stay at home and keep your germs to yourself, thank you very much ... Actually recover, work from home and don't share it with everyone else because no one loves you for sharing ... Decongestants, I think for me.'

Raina MacIntyre, professor of global biosecurity and head of Biosecurity Program at the Kirby Institute, University of New South Wales, Sydney, Australia:

'If I can prevent it, I want to prevent it, so I do wear masks in public spaces. If I go shopping, or whatever, I put on a mask. I wear it at work in our workplace ... In my team, everyone was wearing masks till about 2023. And now I'm the only one ... but we've got a big air purifier in our area where we have meetings, and we have that going with a HEPA filter. So I use air purifiers at home and at work, and I also use antiviral nasal sprays. I've seen clinical studies on sprays containing carrageenan ... and the other one where there has been a clinical trial is ... the nitric oxide spray. [If I get a cold], I would see my GP and try to get antivirals if it was covid or influenza ... so testing is really important. If you don't get tested, you can't get antivirals ... I use those antiviral sprays as treatment as well, because one of the studies of that was on SARS-CoV-2 ... showed that it actually clears the virus faster if you take it, but I have to say, my rate of getting respiratory infections is pretty low since I've started wearing masks everywhere ... If I've got aches and pains, or whatever yeah, I'll avail myself of some Panadol*.'

Leo Poon, professor of public health virology and head of Division of Public Health Laboratory Sciences, University of Hong Kong Li Ka Shing Faculty of Medicine, Hong Kong:

'Probably I'd just wear a mask. I still would go to work ... If I have a severe flu, then I just basically go to bed and take some rest ... just answering emails, right? I try not to spread the disease. If I know I'm very contagious, I'd better just stay at home.'

---

* Brand name for paracetamol.

Julian Tang, consultant virologist at University Hospitals Leicester, UK:

'I get my flu and COVID vaccination each season, partly because I'm a doctor, but also I think that will boost my immunity, and also give me cross-reactive protection against other strains that are evolving all the time ... [when I get a cold,] I just put up with it and blow my nose. The main thing ... is I really try and clear out the infected snot by blowing my nose. I mean, you can phrase that more politely if you want to, but if you sniff it back up, it causes problems. I used to do this, and then got a very bad sinusitis, and secondary bacterial infection. So you've got to get rid of all the infected fluid, whether it's by coughing it out, spitting it out, blowing it out in some way. Because if you don't, you get bacterial colonisation, and then you're in trouble ... I take paracetamol ... if I get a headache, if I get tired, and I get fever.'

Ronald Turner, professor emeritus in the Department of Pediatrics, University of Virginia, Charlottesville, USA:

'I actually fairly vigorously treat. So I use nasal decongestant. Topical. So oxymetazoline or xylometazoline, which are very potent ... For scratchy throat and nasal secretions, rhinorrhea, I use antihistamines, and I tend to use ... diphenhydramine. It's not labelled for common colds. It's labelled for allergy but it's the most potent first-generation antihistamine for stopping runny nose and cough ... I use analgesics if necessary, but primarily nasal decongestant, and antihistamine ... Zinc and all that crap is just that. Crap. I don't. I don't think it does anything.'

Acknowledgements

To write a book, life must have shaped one into the author who could write that book. That's why I'd like to start by thanking all of the friends, family, enemies and casual acquaintances who spent the last few decades shaping me into the author who wrote this book.

Whether they shaped me into an author who could write it well is for the reader to judge but thanks to the efforts of my agent, Eli Keren of Curious Minds, and Canelo's editorial team of Martina O'Sullivan, Hannah Taylor, Abbie Headon and Chere Tricot, it's a considerably better book than I could have produced on my own. It's certainly a better book for the illustrations by Sam Andebonn.

I'd also like to thank any and all of the Chalk Scribblers who have shared your opinions and particularly to Olga Gridina, who keeps it all going. Thank you also to my sister, Katharine Kelly, for applying her expertise in building physics and for making it possible in so many other ways.

The subject matter is drawn from an enormous amount of work carried out by an enormous number of people, some of whom are named in the book but most of whom are not. All of those people share the credit for what we know about colds and, consequently, for this book. I'd like to extend my particular appreciation to the experts who took the time to share their insights and sanity-check mine: Christopher Brown, Henry Burridge, James Cherry, Sheldon Cohen, Amanda Cole, Ronald Eccles, Ron Fouchier, Stephen Griffin, Ruth Harvey, Helen Hayes, Jose-Luis Jimenez, Raina McIntyre, Mark Mon-Williams, Lidia Morawska, Sanchita Pal, Leo Poon, Stefania Renna, Martin Sikora, Ronald Turner, Julian Tang and Gabriela Zapata-Lancaster.

I'd also like to thank the people who shared their experiences of long-term debilitation following infections that most people shrug off. It took courage to be candid and it was very generous to share functional time that they have to ration so carefully. Many thanks to Amina Akhtar, Wendy Heard and Lucy.

And finally, many thanks to all of the friends who have spent the last two years listening to me rambling on about the latest research rabbit hole I've fallen down and, on more than one occasion, hauled me out of it.

If you've found this book useful or enjoyable, all of the above people deserve a share of the credit. If you didn't, the responsibility is entirely my own.

Notes

1. Bosworth, M. L., et al. 2023. "Risk of New-Onset Long COVID Following Reinfection With Severe Acute Respiratory Syndrome Coronavirus 2: A Community-Based Cohort Study." *Open Forum Infectious Diseases* 10: ofad493.

2. Lucy. 2025. Interview by author, January 24, 2025.

3. Ibid.

4. Janicki-Deverts, D., and C. N. Crittenden. 2020. "Common Cold: Cause." In *Encyclopedia of Behavioral Medicine*, edited by M. Gellman, 504–505. Cham, Switzerland: Springer Nature Switzerland AG.

5. Hayes, H., et al. 2024. *Employer Costs from Respiratory Infections: Survey Data on the Business Burden*. London, UK: Office of Health Economics.

6. Xe Currency Converter. 2025. "Historical Rate Tables." Xe Corporation Inc. Accessed November 19, 2025. https://www.xe.com/currencytables/?from=GBP&date=2023-07-01#table-section.

7. Scanes, C. G. 2018. "The Neolithic Revolution, Animal Domestication, and Early Forms of Animal Agriculture." In *Animals and Human Society*, edited by C. Scanes and S. Toukhsati, 103–131. London, UK: Elsevier Inc.

8. Memoli, M. J., et al. 2020. "Influenza A Reinfection in Sequential Human Challenge: Implications for Protective Immunity and 'Universal' Vaccine Development." *Clinical Infectious Diseases* 70: 748–753.

9. Bobrovitz, N., et al. 2023. "Protective Effectiveness of Previous SARS-CoV-2 Infection and Hybrid Immunity Against the Omicron Variant and Severe Disease: A Systematic Review and Meta-Regression." *Lancet Infectious Diseases* 23: 556–567.

10. Nightingale, F. 1863. *Notes on Hospitals*. London, UK: Longman, Green, Longman, Roberts and Green.

11. Allander, T., et al. 2005. "Cloning of a Human Parvovirus by Molecular Screening of Respiratory Tract Samples." *Proceedings of the National Academy of Sciences of the USA* 102: 12891–12896.

12. Dochez, A. R. 1912. "The Occurrence and Virulence of Pneumococci in the Circulating Blood During Lobar Pneumonia and the Susceptibility of Pneumococcus

Strains to Univalent Antipneumococcus Serum." *Journal of Experimental Medicine* 16: 680–692.

13. Dochez, A. R., and S. Benison. 1955. "*Oral History Interview with Alphonse Raymond Dochez, 1955.*" New York: Columbia University. Accessed December 28, 2022. https://dlc.library.columbia.edu/catalog/cul:1c59zw3t2h.

14. Ibid.

15. Heidelberger, M., et al. 1971. *Alphonse Raymond Dochez 1882–1964*. Washington DC: National Academy of Sciences.

16. Ibid.

17. De Kruif, P. 1927. *Microbe Hunters*. London: Jonathan Cape.

18. Tyrrell, D., and M. Fielder. 2002. *Cold Wars: The Fight Against the Common Cold*. Oxford: Oxford University Press.

19. Foster Jr, G. B. 1917. "The Etiology of Common Colds: The Probable Role of a Filtrable Virus as the Causative Factor: With Experiments on the Cultivation of a Minute Micro-Organism from the Nasal Secretion Filtrates." *Journal of Infectious Diseases* 21: 451–474.

20. Shibley, G. S., et al. 1926. "Studies in the Common Cold: I. Observations of the Normal Bacterial Flora of Nose and Throat with Variations Occurring During Colds." *Journal of Experimental Medicine* 43: 415–431.

21. Dochez, A. R., and S. Benison. 1955. "*Oral History Interview with Alphonse Raymond Dochez, 1955.*" New York: Columbia University. Accessed December 28, 2022. https://dlc.library.columbia.edu/catalog/cul:1c59zw3t2h.

22. Dochez, A. R., et al. 1930. "Studies in the Common Cold: IV. Experimental Transmission of the Common Cold to Anthropoid Apes and Human Beings by Means of a Filtrable Agent." *Journal of Experimental Medicine* 52: 701–716.

23. Sender, R., et al. 2016. "Revised Estimates for the Number of Human and Bacteria Cells in the Body." *PLoS Biology* 14: e1002533.

24. Virgin, H. W., et al. 2009. "Redefining Chronic Viral Infection." *Cell* 138: 30–50.

25. Gafafer, W. M., and J. A. Doull. 1933. "Stability of Resistance to the Common Cold." *American Journal of Epidemiology* 18: 712–726.

26. Gafafer, W. M. 1932. "Eye Color and Disease of the Upper Respiratory Tract (Common Cold)." *American Journal of Epidemiology* 16: 880–884.

27. Gafafer, W. M. 1932. "Disease of the Upper Respiratory Tract (Common Cold) in Jews and Non-Jews." *Human Biology* 4: 429–433.

28. Sargent II, F., et al. 1947. "Further Studies on Stability of Resistance to the Common Cold; The Importance of Constitution." *American Journal of Hygiene* 45: 29–32.

29. Medawar, P. B., and J. S. Medawar. 1985. *Aristotle to Zoos: A Philosophical Dictionary of Biology*. Cambridge: Harvard University Press.

30. Eccles, R. 1996. "A Role for the Nasal Cycle in Respiratory Defence." *European Respiratory Journal* 9: 371–376.

31. Bustamante-Marin, X. M., and L. E. Ostrowski. 2017. "Cilia and Mucociliary Clearance." *Cold Spring Harbor Perspectives in Biology* 9: a028241.

32. Eccles, R. 1996. "A Role for the Nasal Cycle in Respiratory Defence." *European Respiratory Journal* 9: 371–376.

33. Naclerio, R. M., et al. 1988. "Kinins Are Generated During Experimental Rhinovirus Colds." *Journal of Infectious Diseases* 157: 133–142.

34. Winther, B., et al. 1984. "Histopathologic Examination and Enumeration of Polymorphonuclear Leukocytes in the Nasal Mucosa During Experimental Rhinovirus Colds." *Acta Oto-Laryngologica* 413: 19–24.

35. Murgia, V., et al. 2020. "Upper Respiratory Tract Infection-Associated Acute Cough and the Urge to Cough: New Insights for Clinical Practice." *Pediatric Allergy, Immunology, and Pulmonology* 33: 3–11.

36. Morice, A. H. 2002. "Epidemiology of Cough." *Pulmonary Pharmacology & Therapeutics* 15: 253–259.

37. Jackson, G. G., et al. 1958. "Transmission of the Common Cold to Volunteers Under Controlled Conditions. I. The Common Cold as a Clinical Entity." *AMA Archives of Internal Medicine* 101: 267–278.

38. Ison, M. G., et al. 2002. "Current Research on Respiratory Viral Infections: Fourth International Symposium." *Antiviral Research* 55: 227–278.

39. Turner, R. B. 2024. Interview by author, November 15, 2025.

40. Naclerio, R. M., et al. 1988. "Kinins Are Generated During Experimental Rhinovirus Colds." *Journal of Infectious Diseases* 157: 133–142.

41. Turner, R. B., et al. 1998. "Association Between Interleukin-8 Concentration in Nasal Secretions and Severity of Symptoms of Experimental Rhinovirus Colds." *Clinical Infectious Diseases* 26: 840–846.

42. Heikkinen, T., and M. D. Järvinen. 2003. "The Common Cold." *Lancet* 361: 51–59.

43. Hendley, J. O. 1998. "Epidemiology, Pathogenesis, and Treatment of the Common Cold." *Seminars in Pediatric Infectious Diseases* 9: 50–55.

44. Sikora, M. 2025. Interview by author, August 8, 2025.

45. Pitulko, V. V., et al. 2004. "The Yana RHS Site: Humans in the Arctic Before the Last Glacial Maximum." *Science* 303: 52–56.

46. Sikora, M., et al. 2019. "The Population History of Northeastern Siberia Since the Pleistocene." *Nature* 570: 182–188.

47. Nielsen, S. H., et al. 2021. "31,600-Year-Old Human Virus Genomes Support a Pleistocene Origin for Common Childhood Infections." *bioRxiv*: 2021.2006.2028.450199.

48. Wolfe, N. D., et al. 2007. "Origins of Major Human Infectious Diseases." *Nature* 447: 279–283.

49. Lynch III, J. P., et al. 2011. "Adenovirus." *Seminars in Respiratory and Critical Care Medicine* 32: 494–511.

50. Dobson, A. P., and E. R. Carper. 1996. "Infectious Diseases and Human Population History: Throughout History the Establishment of Disease Has Been a Side Effect of the Growth of Civilization." *BioScience* 46: 115–126.

51. Hublin, J.-J., et al. 2017. "New Fossils from Jebel Irhoud, Morocco and the Pan-African Origin of Homo Sapiens." *Nature* 546: 289–292.

52. Scanes, C. G. 2018. "The Neolithic Revolution, Animal Domestication, and Early Forms of Animal Agriculture." In *Animals and Human Society*, edited by C. Scanes and S. Toukhsati, 103–131. London, UK: Elsevier Inc.

53. Ibid.

54. Wolfe, N. D., et al. 2007. "Origins of Major Human Infectious Diseases." *Nature* 447: 279–283.

55. Houldcroft, C. J., and S. Underdown. 2023. "Infectious Disease in the Pleistocene: Old Friends or Old Foes?" *American Journal of Biological Anthropology* 182: 513–531.

56. Dobson, A. P., and E. R. Carper. 1996. "Infectious Diseases and Human Population History: Throughout History the Establishment of Disease Has Been a Side Effect of the Growth of Civilization." *BioScience* 46: 115–126.

57. Schiefsky, M. J. 2005. *Hippocrates on Ancient Medicine: Translated with Introduction and Commentary*. Leiden, Netherlands: Koninklijke Brill NV.

58. Witek, T. J., et al. 2015. "The Natural History of Community-Acquired Common Colds Symptoms Assessed Over 4-Years." *Rhinology* 53: 81–88.

59. Creighton, C. 1894. *A History of Epidemics in Britain, Volume 2: From the Extinction of Plague to the Present Time*. Cambridge, UK: Cambridge University Press.

60. Hewer, C. L. 1979. "1918 Influenza Epidemic." *British Medical Journal* 1: 199.

61. Creighton, C. 1894. *A History of Epidemics in Britain, Volume 2: From the Extinction of Plague to the Present Time*. Cambridge, UK: Cambridge University Press.

62. Barberis, I., et al. 2016. "History and Evolution of Influenza Control Through Vaccination: From the First Monovalent Vaccine to Universal Vaccines." *Journal of Preventative Medicine and Hygiene* 57: E115–E120.

63. Creighton, C. 1894. *A History of Epidemics in Britain, Volume 2: From the Extinction of Plague to the Present Time*. Cambridge, UK: Cambridge University Press.

64. Burnet, F. M., and E. Clark. 1942. *Influenza: A Survey of the Last 50 Years in the Light of Modern Work on the Virus of Epidemic Influenza*. Melbourne, Australia: Macmillan.

65. Molineux. 1694. "Dr. Molineux's Historical Account of the Late General Coughs and Colds; with Some Observations on Other Epidemick Distempers." *Philosophical Transactions of the Royal Society (1683-1775)* 18: 105–111.

66. Ibid.

67. Tansey, E. M., and L. A. Reynolds. 1997. "The MRC Common Cold Unit." In *Wellcome Witnesses to Twentieth Century Medicine Volume 2*, edited by E. Tansey et al., 209–267. London, UK: Wellcome Trust.

68. Andrewes, C. H. 1973. *In Pursuit of the Common Cold*. London, UK: Heinemann Medical.

69. Ibid.

70. Editorial. 1963. "William Firth Wells." *The Evening Sun* (Baltimore, USA), October 10, 1963.

71. Wells, W. F. 1934. "On Air-Borne Infection: Study II. Droplets and Droplet Nuclei." *American Journal of Epidemiology* 20: 611–618.

72. Wang, C. C., et al. 2021. "Airborne Transmission of Respiratory Viruses." *Science* 373: eabd9149.

73. Ibid.

74. Riley, R. L. 2001. "What Nobody Needs to Know About Airborne Infection." *American Journal of Respiratory and Critical Care Medicine* 163: 7–8.

75. Wells, W. F., and M. W. Wells. 1936. "Air-Borne Infection: Sanitary Control." *Journal of the American Medical Association* 107: 1805–1809.

76. Wilder, T. S. 1935. "A School Physician Studies Contagious Diseases." *Childhood Education* 11: 199–204.

77. Wells, W. F., et al. 1939. "Infection of Air: Bacteriologic and Epidemiologic Factors." *American Journal of Public Health and the Nation's Health* 29: 863–880.

78. Kowalski, W. 2009. *Ultraviolet Germicidal Irradiation Handbook: UVGI for Air and Surface Disinfection*. Berlin, Germany: Springer-Verlag.

79. Ibid.

80. Wells, W. F., et al. 1942. "The Environmental Control of Epidemic Contagion: I. An Epidemiologic Study of Radiant Disinfection of Air in Day Schools." *American Journal of Epidemiology* 35: 97–121.

81. Anderson, R. M., and R. M. May. 1985. "Vaccination and Herd Immunity to Infectious Diseases." *Nature* 318: 323–329.

82. Schultz, M. G., and W. Schaffner. 2015. "Alexander Duncan Langmuir." *Emerging Infectious Diseases* 21: 1635–1637.

83. Langmuir, A. D. 1951. "The Potentialities of Biological Warfare Against Man – An Epidemiological Appraisal." *Military Surgeon* 108: 429–430.

84. Ibid.

85. Perkins, J. E., et al. 1947. "The Present Status of the Control of Air-Borne Infections." *American Journal of Public Health and the Nation's Health* 37: 13–22.

86. Langmuir, A. D. 1980. "Changing Concepts of Airborne Infection of Acute Contagious Diseases: A Reconsideration of Classic Epidemiologic Theories." *Annals of the New York Academy of Sciences* 353: 35–44.

87. Langmuir, A. D. 1951. "The Potentialities of Biological Warfare Against Man – An Epidemiological Appraisal." *Military Surgeon* 108: 429–430.

88. Randall, K., et al. 2021. "How Did We Get Here: What Are Droplets and Aerosols and How Far Do They Go? A Historical Perspective on the Transmission of Respiratory Infectious Diseases." *Interface Focus* 11: 20210049.

89. Riley, R. L. 2001. "What Nobody Needs to Know About Airborne Infection." *American Journal of Respiratory and Critical Care Medicine* 163: 7–8.

90. Ibid.

91. Riley, R. L., et al. 1962. "Infectiousness of Air From a Tuberculosis Ward. Ultraviolet Irradiation of Infected Air: Comparative Infectiousness of Different Patients." *American Review of Respiratory Disease* 85: 511–525.

92. Lovelock, J. 2000. *Homage to Gaia: The Life of an Independent Scientist*. Oxford, UK: Oxford University Press.

93. National Research Council (US) Committee on Air Quality in Passenger Cabins of Commercial Aircraft. 2002. *The Airliner Cabin Environment and the Health of Passengers and Crew*. Washington DC, USA: National Academies Press.

94. Lovelock, J. 2000. *Homage to Gaia: The Life of an Independent Scientist*. Oxford, UK: Oxford University Press.

95. Tyrrell, D., and M. Fielder. 2002. *Cold Wars: The Fight Against the Common Cold.* Oxford: Oxford University Press.

96. Andrewes, C. H. 1973. *In Pursuit of the Common Cold.* London, UK: Heinemann Medical.

97. Lovelock, J. 2000. *Homage to Gaia: The Life of an Independent Scientist.* Oxford, UK: Oxford University Press.

98. Tyrrell, D., and M. Fielder. 2002. *Cold Wars: The Fight Against the Common Cold.* Oxford: Oxford University Press.

99. Lovelock, J. 2000. *Homage to Gaia: The Life of an Independent Scientist.* Oxford, UK: Oxford University Press.

100. Lovelock, J. E., et al. 1952. "Further Studies on the Natural Transmission of the Common Cold." *Lancet* 260: 657–660.

101. Hendley, J. O. 1998. "Epidemiology, Pathogenesis, and Treatment of the Common Cold." *Seminars in Pediatric Infectious Diseases* 9: 50–55.

102. Lovelock, J. 2000. *Homage to Gaia: The Life of an Independent Scientist.* Oxford, UK: Oxford University Press.

103. Wat, D. 2004. "The Common Cold: A Review of the Literature." *European Journal of Internal Medicine* 15: 79–88.

104. Hendley, J. O., et al. 1973. "Transmission of Rhinovirus Colds by Self-Inoculation." *New England Journal of Medicine* 288: 1361–1364.

105. Gwaltney Jr, J. M., et al. 1978. "Hand-to-Hand Transmission of Rhinovirus Colds." *Annals of Internal Medicine* 88: 463–467.

106. Dick, E. C., et al. 1987. "Aerosol Transmission of Rhinovirus Colds." *Journal of Infectious Diseases* 156: 442–448.

107. Ibid.

108. Langmuir, A. D. 1980. "Changing Concepts of Airborne Infection of Acute Contagious Diseases." *Annals of the New York Academy of Sciences* 353: 35–44.

109. Xie, X., et al. 2007. "How Far Droplets Can Move in Indoor Environments–Revisiting the Wells Evaporation-Falling Curve." *Indoor Air* 17: 211–225.

110. Greenhalgh, T., et al. 2022. "How Covid-19 Spreads: Narratives, Counter Narratives, and Social Dramas." *British Medical Journal* 378: e069940.

111. Molteni, M. 2021. "The 60-Year-Old Scientific Screwup That Helped Covid Kill." *Wired.* Accessed December 28, 2022. https://www.wired.com/story/the-teeny-tiny-scientific-screwup-that-helped-covid-kill/.

112. Brown, H. M., and K. R. Irving. 1973. "The Size and Weight of Common Allergenic Pollens." *Acta Allergologica* 28: 132–137.

113. Jimenez, J. L. 2025. Interview by author, September 24, 2025.

114. Ibid.

115. Petersen, E., et al. 2020. "Comparing SARS-CoV-2 with SARS-CoV and Influenza Pandemics." *Lancet Infectious Diseases* 20: e238–e244.

116. Molteni, M. 2021. "The 60-Year-Old Scientific Screwup That Helped Covid Kill." *Wired*.

117. Randall, K. 2022. "*The Tiny COVID Mistake with Deadly Implications.*" TEDXKC 2022 Kansas City.

118. Randall, K., et al. 2021. "How Did We Get Here: What Are Droplets and Aerosols and How Far Do They Go? A Historical Perspective on the Transmission of Respiratory Infectious Diseases." *Interface Focus* 11: 20210049.

119. Jimenez, J. L., et al. 2022. "What Were the Historical Reasons for the Resistance to Recognizing Airborne Transmission During the COVID-19 Pandemic?" *Indoor Air* 32: e13070.

120. Greenhalgh, T., et al. 2022. "How Covid-19 Spreads: Narratives, Counter Narratives, and Social Dramas." *British Medical Journal* 378: e069940.

121. Kupferschmidt, K. 2022. "WHO's Departing Chief Scientist Regrets Errors in Debate Over Whether SARS-CoV-2 Spreads Through Air." *Science*. Accessed May 25, 2025. https://www.science.org/content/article/who-s-departing-chief-scientist-regrets-errors-debate-over-whether-sars-cov-2-spreads#.Y3-Ofih6nDI.linkedin.

122. Murray, J. 2025. "Nurseries in England Bring in Covid-Style Protocols as Measles Cases Rise." *The Guardian*, July 19, 2025.

123. Collins, W. 1859. *The Woman in White*. Duke Classics.

124. Baer, R. D., et al. 2008. "Cross-Cultural Perspectives on Physician and Lay Models of the Common Cold." *Medical Anthropology Quarterly* 22: 148–166.

125. Helman, C. G. 1978. "'Feed a Cold, Starve a Fever' – Folk Models of Infection in an English Suburban Community, and Their Relation to Medical Treatment." *Culture, Medicine and Psychiatry* 2: 107–137.

126. Sargent II, F., et al. 1947. "Further Studies on Stability of Resistance to the Common Cold; The Importance of Constitution." *American Journal of Hygiene* 45: 29–32.

127. Ibid.

128. Andrewes, C. H. 1949. "The Natural History of the Common Cold." *Lancet* 253: 71–75.

129. Tyrrell, D., and M. Fielder. 2002. *Cold Wars: The Fight Against the Common Cold.* Oxford: Oxford University Press.

130. Mudd, S., and S. B. Grant. 1919. "Reactions to Chilling of the Body Surface: Experimental Study of a Possible Mechanism for the Excitation of Infections of the Pharynx and Tonsils." *Journal of Medical Research* 40: 53–101.

131. Cruz, A. A., and A. Togias. 2008. "Upper Airways Reactions to Cold Air." *Current Allergy and Asthma Reports* 8: 111–117.

132. Andrewes, C. H. 1950. "Adventures Among Viruses; The Puzzle of the Common Cold." *New England Journal of Medicine* 242: 235–240.

133. Ibid.

134. Jackson, G. G. 1989. "The Medical Research Council Common Cold Unit, Harvard Hospital, Salisbury, Wilts, England." *Reviews of Infectious Diseases* 11: 1020–1021.

135. Dowling, H. F., et al. 1958. "Transmission of the Common Cold to Volunteers Under Controlled Conditions. III. The Effect of Chilling of the Subjects Upon Susceptibility." *American Journal of Hygiene* 68: 59–65.

136. Andrewes, C. H. 1953. "The Common Cold." *British Medical Bulletin* 9: 206–207.

137. Eccles, R. 2025. Interview by author, March 13, 2025.

138. Ibid.

139. Epps, G. 1982. "Stalking the Common Cold." *New York Times Magazine*, November 14, 1982.

140. Eccles, R. 2025. Interview by author, March 13, 2025.

141. Johnson, C., and R. Eccles. 2005. "Acute Cooling of the Feet and the Onset of Common Cold Symptoms." *Family Practice* 22: 608–613.

142. Inada, M. 2021. "Post-Covid-19, World Risks Having to Pay Off 'Immunity Debt'." *Wall Street Journal*, June 28, 2021.

143. Jarry, J. 2022. "Claims of an Immunity Debt in Children Owe Us Evidence." McGill University. Accessed July 30, 2025. https://www.mcgill.ca/oss/article/covid-19-medical-critical-thinking/claims-immunity-debt-children-owe-us-evidence.

144. French Society of Paediatrics.

145. Cohen, R., et al. 2021. "Pediatric Infectious Disease Group (GPIP) Position Paper on the Immune Debt of the COVID-19 Pandemic in Childhood, How Can We Fill the Immunity Gap?" *Infectious Diseases Now* 51: 418–423.

146. Gupta, S. 2022. "Lockdowns Put Us at the Mercy of Disease." *The Telegraph*, December 9, 2022.

147. Moriyama, M., et al. 2020. "Seasonality of Respiratory Viral Infections." *Annual Review of Virology* 7: 83–101.

148. Yu, H., et al. 2013. "Characterization of Regional Influenza Seasonality Patterns in China and Implications for Vaccination Strategies: Spatio-Temporal Modeling of Surveillance Data." *PLoS Medicine* 10: e1001552.

149. Cohen, R., et al. 2023. "Immune Debt: Recrudescence of Disease and Confirmation of a Contested Concept." *Infectious Diseases Now* 53: 104638.

150. Netea, M. G., et al. 2020. "Defining Trained Immunity and its Role in Health and Disease." *Nature Reviews Immunology* 20: 375–388.

151. Andrewes, C. H. 1965. *The Common Cold*. London, UK: Weidenfeld and Nicolson.

152. Hendley, J. O. 1998. "Epidemiology, Pathogenesis, and Treatment of the Common Cold." *Seminars in Pediatric Infectious Diseases* 9: 50–55.

153. Lovelock, J. 2000. *Homage to Gaia: The Life of an Independent Scientist*. Oxford, UK: Oxford University Press.

154. Andrewes, C. H., et al. 1951. "An Experiment on the Transmission of Colds." *Lancet* 257: 25–27.

155. Dick, E. C., et al. 1980. "Possible Modification of the Normal Winter Fly-In Respiratory Disease Outbreak at McMurdo Station." *Antarctic Journal of the United States* 15: 173–174.

156. Dick, E. C., et al. 1978. "Respiratory Virus Transmission at McMurdo Station and Scott Base (New Zealand) During the Winter Fly-In Period." *Antarctic Journal of the United States* 13: 170–171.

157. Dick, E. C., et al. 1967. "Epidemiology of Infections with Rhinovirus Types 43 and 55 in a Group of University of Wisconsin Student Families." *American Journal of Epidemiology* 86: 386–400.

158. Cluff, L. E., et al. 1966. "Asian Influenza. Infection, Disease, and Psychological Factors." *Archives of Internal Medicine* 117: 159–163.

159. Totman, R., et al. 1980. "Predicting Experimental Colds in Volunteers From Different Measures of Recent Life Stress." *Journal of Psychosomatic Research* 24: 155–163.

160. Cohen, S., et al. 1991. "Psychological Stress and Susceptibility to the Common Cold." *New England Journal of Medicine* 325: 606–612.

161. Cohen, S., et al. 1997. "Social Ties and Susceptibility to the Common Cold." *Journal of the American Medical Association* 277: 1940–1944.

162. Cohen, S., et al. 2008. "Objective and Subjective Socioeconomic Status and Susceptibility to the Common Cold." *Health Psychology* 27: 268–274.

163. Miller, G. E., et al. 2002. "Chronic Psychological Stress and the Regulation of Pro-Inflammatory Cytokines: A Glucocorticoid-Resistance Model." *Health Psychology* 21: 531–541.

164. Cohen, S., et al. 2012. "Chronic Stress, Glucocorticoid Receptor Resistance, Inflammation, and Disease Risk." *Proceedings of the National Academy of Sciences of the USA* 109: 5995–5999.

165. Lamers, M. M., and B. L. Haagmans. 2022. "SARS-CoV-2 Pathogenesis." *Nature Reviews Microbiology* 20: 270–284.

166. Davis, H. E., et al. 2023. "Long COVID: Major Findings, Mechanisms and Recommendations." *Nature Reviews Microbiology* 21: 133–146.

167. Tansey, E. M., and L. A. Reynolds. 1997. "The MRC Common Cold Unit." In *Wellcome Witnesses to Twentieth Century Medicine Volume 2*, edited by E. Tansey et al., 209–267. London, UK: Wellcome Trust.

168. Lovelock, J. 2000. *Homage to Gaia: The Life of an Independent Scientist*. Oxford, UK: Oxford University Press.

169. Lidwell, O. M., and T. Sommerville. 1951. "Observations on the Incidence and Distribution of the Common Cold in a Rural Community During 1948 and 1949." *Journal of Hygiene* 49: 365–381.

170. Department of Economic and Social Affairs. 2019. *World Urbanization Prospects: The 2018 Revision (ST/ESA/SER.A/420)*. New York, USA: United Nations.

171. Lidwell, O. M., and R. E. O. Williams. 1961. "The Epidemiology of the Common Cold. I." *Journal of Hygiene* 59: 309–319.

172. WHO. 2013. *Pandemic Influenza Risk Management. WHO Interim Guidance*. Geneva: WHO.

173. Ibid.

174. Moriyama, M., et al. 2020. "Seasonality of Respiratory Viral Infections." *Annual Review of Virology* 7: 83–101.

175. Suryadevara, M., and J. B. Domachowske. 2021. "Epidemiology and Seasonality of Childhood Respiratory Syncytial Virus Infections in the Tropics." *Viruses* 13: 696.

176. Yu, H., et al. 2013. "Characterization of Regional Influenza Seasonality Patterns in China and Implications for Vaccination Strategies: Spatio-Temporal Modeling of Surveillance Data." *PLoS Medicine* 10: e1001552.

177. Moriyama, M., et al. 2020. "Seasonality of Respiratory Viral Infections." *Annual Review of Virology* 7: 83–101.

178. Xie, X., et al. 2007. "How Far Droplets Can Move in Indoor Environments– Revisiting the Wells Evaporation-Falling Curve." *Indoor Air* 17: 211–225.

179. CIBSE. 2020. *TM40: Health and Wellbeing in Building Services.* London, UK: CIBSE.

180. Pleil, J. D., et al. 2021. "The Physics of Human Breathing: Flow, Timing, Volume, and Pressure Parameters for Normal, On-Demand, and Ventilator Respiration." *Journal of Breath Research* 15: 10.1088/1752-7163/ac2589.

181. Moriyama, M., et al. 2020. "Seasonality of Respiratory Viral Infections." *Annual Review of Virology* 7: 83–101.

182. Sale, C. S. 1972. "Humidification to Reduce Respiratory Illnesses in Nursery School Children." *Southern Medical Journal* 65: 882–885.

183. ONS. 2023. "Dataset: Workforce and School Absence Because of Health Reasons." Office for National Statistics. Accessed August 1, 2025. https://www.ons.gov.uk/peoplepopulationandcommunity/healthandsocialcare/healthandwellbeing/datasets/workforceandschoolabsencebecauseofhealthreasons.

184. Viitanen, H., et al. 2011. "Mould Growth Modelling to Evaluate Durability of Materials." *International Conference on Durability of Building Materials and Components: Conference Proceedings* XII DBMC: 409–416.

185. Fendrick, A. M., et al. 2003. "The Economic Burden of Non-Influenza-Related Viral Respiratory Tract Infection in the United States." *Archives of Internal Medicine* 163: 487–494.

186. Molinari, N.-A. M., et al. 2007. "The Annual Impact of Seasonal Influenza in the US: Measuring Disease Burden and Costs." *Vaccine* 25: 5086–5096.

187. World Bank. 2025. "Population, Total – United States." World Bank Group. Accessed August 1, 2025. https://data.worldbank.org/indicator/SP.POP.TOTL?locations=US.

188. Bureau of Labor Statistics. 2025. "CPI Inflation Calculator." Bureau of Labor Statistics. Accessed August 1, 2025. https://data.bls.gov/cgi-bin/cpicalc.pl?cost1=1&year1=200306&year2=202506.

189. Xe Currency Converter. 2025. "Historical Rate Tables." Xe Corporation Inc. Accessed September 5, 2025. https://www.xe.com/currencytables/?from=GBP&date=2025-07-01#table-section.

190. Hayes, H., et al. 2024. *Employer Costs from Respiratory Infections: Survey Data on the Business Burden.* London, UK: Office of Health Economics.

191. Office for Budget Responsibility. 2025. *A Brief Guide to the UK Public Finances.* London, UK: Office for Budget Responsibility.

192. Xe Currency Converter. 2025. "Historical Rate Tables." Xe Corporation Inc. Accessed September 3, 2025. https://www.xe.com/currencytables/?from=GBP&date=2025-09-01#table-section.

193. Siddique, H. 2024. "Presenteeism: What is Causing Britain's Working-While-Sick Epidemic?" *The Guardian*. Accessed May 19, 2025. https://www.theguardian.com/society/article/2024/aug/02/presenteeism-what-causing-britain-working-while-sick-epidemic.

194. O'Halloran, J., and C. Thomas. 2024. *Healthy Industry, Prosperous Economy*. London, UK: Institute for Public Policy Research.

195. Johnson, N. P. A. S., and J. Mueller. 2002. "Updating the Accounts: Global Mortality of the 1918-1920 'Spanish' Influenza Pandemic." *Bulletin of the History of Medicine* 76: 105–115.

196. McKnight, M. 2020. "Into the Wild. Twice. For Mankind." *Sports Illustrated*, May 27, 2020.

197. Stern, M. J. 2012. "The Worst Pandemic in History." *Slate*, December 26, 2012.

198. Rozell, N. 1998. "Permafrost Preserves Clues to Deadly 1918 Flu." Geophysical Institute, University of Alaska Fairbanks. Accessed August 16, 2024. https://www.gi.alaska.edu/alaska-science-forum/permafrost-preserves-clues-deadly-1918-flu.

199. Taubenberger, J. K., et al. 2007. "Discovery and Characterization of the 1918 Pandemic Influenza Virus in Historical Context." *Antiviral Therapy* 12: 581–591.

200. Racaniello, V. R. 2022. "*TWiV 966: 1918 Influenza with Jeffery Taubenberger.*" Columbia University.

201. Leung, N. H. L., et al. 2015. "Review Article: The Fraction of Influenza Virus Infections That Are Asymptomatic: A Systematic Review and Meta-Analysis." *Epidemiology* 26: 862–872.

202. Wat, D. 2004. "The Common Cold: A Review of the Literature." *European Journal of Internal Medicine* 15: 79–88.

203. Pappas, D. E. 2017. "The Common Cold." In *Principles and Practice of Pediatric Infectious Diseases*: 199–202.e191.

204. Bustamante-Marin, X. M., and L. E. Ostrowski. 2017. "Cilia and Mucociliary Clearance." *Cold Spring Harbor Perspectives in Biology* 9: a028241.

205. Poehling, K. A., et al. 2006. "The Underrecognized Burden of Influenza in Young Children." *New England Journal of Medicine* 355: 31–40.

206. Wright, P. F., et al. 2013. "Orthomyxoviruses." In *Fields Virology*, edited by D. Knipe and P. Howley, 1187–1239. Philadelphia: Wolters Kluwer Health/Lippincott Williams & Wilkins.

207. Rello, J., and A. Pop-Vicas. 2009. "Clinical Review: Primary Influenza Viral Pneumonia." *Critical Care* 13: 235.

208. Hayden, F. G., and P. Palese. 2017. "Influenza Virus." In *Clinical Virology, Fourth Edition*, edited by D. Richman et al., 1009–1058. Washington DC: ASM Press.

209. Barry, J. M. 2004. "The Site of Origin of the 1918 Influenza Pandemic and its Public Health Implications." *Journal of Translational Medicine* 2: 1–4.

210. Taubenberger, J. K., and D. M. Morens. 2006. "1918 Influenza: The Mother of All Pandemics." *Emerging Infectious Diseases* 12: 15–22.

211. Johnson, N. 2006. *Britain and the 1918-19 Influenza Pandemic: A Dark Epilogue*. London, UK: Routledge.

212. Prost, A. 2014. "War Losses." Freie Universität Berlin. Accessed August 7, 2025. https://encyclopedia.1914-1918-online.net/article/war-losses/.

213. Patterson, K. D., and G. F. Pyle. 1991. "The Geography and Mortality of the 1918 Influenza Pandemic." *Bulletin of the History of Medicine* 65: 4–21.

214. Ibid.

215. Toynbee, A. J. 1957. *A Study of History. Abridgement of Volumes VII-X*. Oxford, UK: Oxford University Press.

216. Beiner, G. 2022. "Introduction: The Great Flu Between Remembering and Forgetting." In *Pandemic Re-Awakenings: The Forgotten and Unforgotten 'Spanish' Flu of 1918-1919*, edited by G. Beiner, 1–48. Oxford, UK: Oxford University Press.

217. Collier, R. 1974. *The Plague of the Spanish Lady: The Influenza Pandemic of 1918-1919*. London, UK: Macmillan.

218. Cunha, B. A. 2004. "Influenza: Historical Aspects of Epidemics and Pandemics." *Infectious Disease Clinics of North America* 18: 141–155.

219. Pfeiffer, R. 1893. "Die Aetiologie der Influenza." *Zeitschrift für Hygiene* 13: 357–386.

220. Burnet, F. M., and E. Clark. 1942. *Influenza: A Survey of the Last 50 Years in the Light of Modern Work on the Virus of Epidemic Influenza*. Melbourne, Australia: Macmillan.

221. Osler, W. 1912. *The Principles and Practice of Medicine*. New York, USA: D. Appleton and Company.

222. Newsholme, A. 1919. *Discussion on Influenza*. London, UK: Royal Society of Medicine.

223. Taubenberger, J. K., et al. 2007. "Discovery and Characterization of the 1918 Pandemic Influenza Virus in Historical Context." *Antiviral Therapy* 12: 581–591.

224. Bresalier, M. 2012. "Uses of a Pandemic: Forging the Identities of Influenza and Virus Research in Interwar Britain." *Social History of Medicine* 25: 400–424.

225. Yamanouchi, T., et al. 1919. "The Infecting Agent in Influenza: An Experimental Research." *Lancet* 193: 971.

226. Evans, D. G. 1966. "Wilson Smith, 1897-1965." *Biographical Memoirs of the Fellows of the Royal Society* 12: 478–487.

227. Tyrrell, D. A. J. 1991. "Christopher Howard Andrewes, 7 June 1896–31 December 1987." *Biographical Memoirs of Fellows of the Royal Society* 37: 33–54.

228. Andrewes, C. H. 1965. *The Common Cold*. London, UK: Weidenfeld and Nicolson.

229. Smith, W., et al. 1933. "A Virus Obtained from Influenza Patients." *Lancet* 222: 66–68.

230. Maher, J. A., and J. DeStefano. 2004. "The Ferret: An Animal Model to Study Influenza Virus." *Lab Animal* 33: 50–53.

231. Bresalier, M. 2013. "80 Years Ago Today: MRC Researchers Discover Viral Cause of Flu." *Guardian*, July 8, 2013.

232. Burnet, F. M., and E. Clark. 1942. *Influenza: A Survey of the Last 50 Years*. Melbourne, Australia: Macmillan.

233. Burnet, F. M. 1940. "Influenza Virus Infections of the Chick Embryo Lung." *British Journal of Experimental Pathology* 21: 147–153.

234. Francis, T., et al. 1945. "Protective Effect of Vaccination Against Induced Influenza A." *Journal of Clinical Investigation* 24: 536–546.

235. Salk, J. E., and P. C. Suriano. 1949. "Importance of Antigenic Composition of Influenza Virus Vaccine in Protecting Against the Natural Disease; Observations During the Winter of 1947-1948." *American Journal of Public Health and the Nation's Health* 39: 345–355.

236. Andrewes, C. H. 1950. "Adventures Among Viruses; The Puzzle of the Common Cold." *New England Journal of Medicine* 242: 235–240.

237. Ziegler, T., et al. 2018. "65 Years of Influenza Surveillance by a World Health Organization-Coordinated Global Network." *Influenza and Other Respiratory Viruses* 12: 558–565.

238. Harvey, R. 2025. Interview by author, January 9, 2025.

239. WHO. 2022. "Global Influenza Surveillance and Response System (GISRS)." World Health Organization. Accessed August 17, 2024. https://www.who.int/initiatives/global-influenza-surveillance-and-response-system.

240. WHO. 2013. *Pandemic Influenza Risk Management. WHO Interim Guidance*. Geneva: WHO.

241. Taubenberger, J. K., and D. M. Morens. 2008. "The Pathology of Influenza Virus Infections." *Annual Review of Pathology* 3: 499–522.

242. Ibid.

243. Scholtissek, C., et al. 1985. "The Nucleoprotein as a Possible Major Factor in Determining Host Specificity of Influenza H3N2 Viruses." *Virology* 147: 287–294.

244. Biggerstaff, M., et al. 2014. "Estimates of the Reproduction Number for Seasonal, Pandemic, and Zoonotic Influenza: A Systematic Review of the Literature." *BMC Infectious Diseases* 14: 480.

245. Palese, P. 2004. "Influenza: Old and New Threats." *Nature Medicine* 10: S82–87.

246. Taubenberger, J. K., and D. M. Morens. 2006. "1918 Influenza: The Mother of All Pandemics." *Emerging Infectious Diseases* 12: 15–22.

247. Liu, H., et al. 2024. "Effect of Human H3N2 Influenza Virus Reassortment on Influenza Incidence and Severity During the 2017-18 Influenza Season in the USA: A Retrospective Observational Genomic Analysis." *Lancet Microbe* 5: 100852.

248. Memoli, M. J., et al. 2020. "Influenza A Reinfection in Sequential Human Challenge." *Clinical Infectious Diseases* 70: 748–753.

249. Pleil, J. D., et al. 2021. "The Physics of Human Breathing: Flow, Timing, Volume, and Pressure Parameters for Normal, On-Demand, and Ventilator Respiration." *Journal of Breath Research* 15: 10.1088/1752–7163/ac2589.

250. Kung, H. C., et al. 1978. "Influenza in China in 1977: Recurrence of Influenzavirus A Subtype H1N1." *Bulletin of the World Health Organization* 56: 913–918.

251. Kilbourne, E. D. 2006. "Influenza Pandemics of the 20th Century." *Emerging Infectious Diseases* 12: 9–14.

252. Palese, P. 2004. "Influenza: Old and New Threats." *Nature Medicine* 10: S82–87.

253. Kung, H. C., et al. 1978. "Influenza in China in 1977." *Bulletin of the World Health Organization* 56: 913–918.

254. Burke, D. S., and A. Schleunes. 2024. "A Self-Fulfilling Prophecy Pandemic: The 1977 'Russian Flu'." *Perspectives in Biology and Medicine* 67: 386–405.

255. Fouchier, R. A. 2024. Interview by author, October 30, 2024.

256. Rozo, M., and G. K. Gronvall. 2015. "The Reemergent 1977 H1N1 Strain and the Gain-of-Function Debate." *mBio* 6: e01013–01015.

257. Wright, P. F., et al. 2013. "Orthomyxoviruses." In *Fields Virology*, edited by D. Knipe and P. Howley, 1187–1239. Philadelphia: Wolters Kluwer Health.

258. Dawood, F. S., et al. 2009. "Emergence of a Novel Swine-Origin Influenza A (H1N1) Virus in Humans." *New England Journal of Medicine* 360: 2605–2615.

259. WHO. 2013. *Pandemic Influenza Risk Management. WHO Interim Guidance.* Geneva: WHO.

260. Taubenberger, J. K., et al. 2019. "The 1918 Influenza Pandemic: 100 Years of Questions Answered and Unanswered." *Science Translational Medicine* 11: eaau5485.

261. Taubenberger, J. K., et al. 1997. "Initial Genetic Characterization of the 1918 'Spanish' Influenza Virus." *Science* 275: 1793–1796.

262. Le Bon, L. 1985. "Asia, Pakistan, Batura Glacier Expedition." American Alpine Club. Accessed August 17, 2024. https://publications.americanalpineclub.org/articles/12198533100/Asia-Pakistan-Batura-Glacier-Exploration.

263. McKnight, M. 2020. "Into the Wild. Twice. For Mankind." *Sports Illustrated*, May 27, 2020.

264. Racaniello, V. R. 2022. "*TWiV 966: 1918 Influenza with Jeffery Taubenberger.*" Columbia University.

265. Taubenberger, J. K., et al. 2007. "Discovery and Characterization of the 1918 Pandemic Influenza Virus in Historical Context." *Antiviral Therapy* 12: 581–591.

266. Taubenberger, J. K., and D. M. Morens. 2006. "1918 Influenza: The Mother of All Pandemics." *Emerging Infectious Diseases* 12: 15–22.

267. Fernandez, E. 2002. "The Virus Detective / Dr. John Hultin Has Found Evidence of the 1918 Flu Epidemic That Had Eluded Experts for Decades." *San Francisco Chronicle*, February 17, 2002.

268. Hayward, A. C., et al. 2014. "Comparative Community Burden and Severity of Seasonal and Pandemic Influenza: Results of the Flu Watch Cohort Study." *Lancet Respiratory Medicine* 2: 445–454.

269. Matias, G., et al. 2016. "Modelling Estimates of Age-Specific Influenza-Related Hospitalisation and Mortality in the United Kingdom." *BMC Public Health* 16: 481.

270. ONS. 2024. "Population Estimates for the UK, England, Wales, Scotland, and Northern Ireland: Mid-2022." UK Government. Accessed August 10, 2025. https://www.ons.gov.uk/peoplepopulationandcommunity/populationandmigration/populationestimates/bulletins/annualmidyearpopulationestimates/mid2022.

271. UKHSA. 2025. "Influenza in the UK, Annual Epidemiological Report: Winter 2024 to 2025." UK Government. Accessed August 8, 2025. https://www.gov.uk/government/statistics/influenza-in-the-uk-annual-epidemiological-report-winter-2024-to-2025.

272. Willison, H. J., et al. 2016. "Guillain-Barré Syndrome." *Lancet* 388: 717–727.

273. Sellers, S. A., et al. 2017. "The Hidden Burden of Influenza: A Review of the Extra-Pulmonary Complications of Influenza Infection." *Influenza and Other Respiratory Viruses* 11: 372–393.

274. Leung, N. H. L., et al. 2015. "Review Article: The Fraction of Influenza Virus Infections That Are Asymptomatic: A Systematic Review and Meta-Analysis." *Epidemiology* 26: 862–872.

275. Demicheli, V., et al. 2018. "Vaccines for Preventing Influenza in the Elderly." *Cochrane Database Systematic Reviews* 2: CD004876; Demicheli, V., et al. 2018. "Vaccines for Preventing Influenza in Healthy Adults." *Cochrane Database of Systematic Reviews* 2: CD001269; Jefferson, T., et al. 2018. "Vaccines for Preventing Influenza in Healthy Children." *Cochrane Database of Systematic Reviews* 2: CD004879.

276. Pebody, R., et al. 2019. "End of Season Influenza Vaccine Effectiveness in Adults and Children in the United Kingdom in 2017/18." *EuroSurveillance* 24: 1800488; Pebody, R., et al. 2016. "Effectiveness of Seasonal Influenza Vaccine for Adults and Children in Preventing Laboratory-Confirmed Influenza in Primary Care in the United Kingdom: 2015/16 End-of-Season Results." *EuroSurveillance* 21: 30348; Pebody, R., et al. 2017. "End-of-Season Influenza Vaccine Effectiveness in Adults and Children, United Kingdom, 2016/17." *EuroSurveillance* 22: 17–00306; PHE. 2020. *Surveillance of Influenza and Other Respiratory Viruses in the UK: Winter 2019 to 2020.* London: PHE. 69pp; Pebody, R. G., et al. 2020. "End of Season Influenza Vaccine Effectiveness in Primary Care in Adults and Children in the United Kingdom in 2018/19." *Vaccine* 38: 489–497.

277. Pebody, R., et al. 2019. "End of Season Influenza Vaccine Effectiveness in Adults and Children in the United Kingdom in 2017/18." *EuroSurveillance* 24: 1800488.

278. Krammer, F., et al. 2018. "Influenza." *Nature Reviews Disease Primers* 4: 3.

279. Jefferson, T., et al. 2014. "Neuraminidase Inhibitors for Preventing and Treating Influenza in Adults and Children." *Cochrane Database of Systematic Reviews* 2014: CD008965.

280. Tejada, S., et al. 2021. "Neuraminidase Inhibitors Are Effective and Safe in Reducing Influenza Complications: Meta-Analysis of Randomized Controlled Trials." *European Journal of Internal Medicine* 86: 54–65.

281. McKnight, M. 2020. "Into the Wild. Twice. For Mankind." *Sports Illustrated*, May 27, 2020.

282. Skloot, R. 2010. *The Immortal Life of Henrietta Lacks*. New York: Random House.

283. Andrewes, C. H. 1965. *The Common Cold*. London, UK: Weidenfeld and Nicolson.

284. Tyrrell, D., and M. Fielder. 2002. *Cold Wars: The Fight Against the Common Cold*. Oxford: Oxford University Press.

285. Ibid.

286. Andrewes, C. H. 1973. *In Pursuit of the Common Cold*. London, UK: Heinemann Medical.

287. Tyrrell, D., and M. Fielder. 2002. *Cold Wars: The Fight Against the Common Cold*. Oxford: Oxford University Press.

288. Epps, G. 1982. "Stalking the Common Cold." *New York Times Magazine*, November 14, 1982.

289. Tyrrell, D., and M. Fielder. 2002. *Cold Wars: The Fight Against the Common Cold*. Oxford: Oxford University Press.

290. Cherry, J. D. 2021. Interview by author, April 15, 2021.

291. Lovelock, J. 2000. *Homage to Gaia: The Life of an Independent Scientist*. Oxford, UK: Oxford University Press.

292. Ibid.

293. White, P. R. 1955. *The Cultivation of Animal and Plant Cells*. London: Thames & Hudson.

294. Tyrrell, D., and M. Fielder. 2002. *Cold Wars: The Fight Against the Common Cold*. Oxford: Oxford University Press.

295. Andrewes, C. H. 1953. "The Common Cold." *British Medical Bulletin* 9: 206–207.

296. Hanks, J. H., and F. B. Bang. 1971. "Dr. George Otto Gey 1899-1970." *In Vitro* 6: 3–4.

297. Skloot, R. 2010. *The Immortal Life of Henrietta Lacks*. New York: Random House.

298. Wadman, M. 2017. *The Vaccine Race*. London: Transworld Publishers.

299. Gey, G. O., et al. 1952. "Tissue Culture Studies of the Proliferative Capacity of Cervical Carcinoma and Normal Epithelium." *Cancer Research* 12: 264–265.

300. Hanks, J. H., and F. B. Bang. 1971. "Dr. George Otto Gey 1899-1970." *In Vitro* 6: 3–4.

301. Hilleman, M. R., and J. H. Werner. 1954. "Recovery of New Agent from Patients with Acute Respiratory Illness." *Proceedings of the Society of Experimental Biology and Medicine* 85: 183–188.

302. Collins, H. 1999. "The Man Who Saved Your Life – Maurice R. Hilleman – Developer of Vaccines for Mumps and Pandemic Flu." *Philadelphia Inquirer*, August 30, 1999.

303. Rowe, W. P., et al. 1953. "Isolation of a Cytopathogenic Agent from Human Adenoids Undergoing Spontaneous Degeneration in Tissue Culture." *Proceedings of the Society of Experimental Biology and Medicine* 84: 570–573.

304. Enders, J. F., et al. 1956. "Adenoviruses: Group Name Proposed for New Respiratory-Tract Viruses." *Science* 124: 119–120.

305. Dhingra, A., et al. 2019. "Molecular Evolution of Human Adenovirus (HAdV) Species C." *Scientific Reports* 9: 1039.

306. Lynch III, J. P., et al. 2011. "Adenovirus." *Seminars in Respiratory and Critical Care Medicine* 32: 494–511.

307. Hilleman, M. R., et al. 1957. "Appraisal of Occurrence of Adenovirus-Caused Respiratory Illness in Military Populations." *American Journal of Hygiene* 66: 29–41.

308. Kerr, J. R., and D. Taylor-Robinson. 2007. "David Arthur John Tyrrell CBE: 19 June 1925–2 May 2005." *Biographical Memoirs of Fellows of the Royal Society* 53: 349–363.

309. Tyrrell, D., and M. Fielder. 2002. *Cold Wars: The Fight Against the Common Cold*. Oxford: Oxford University Press.

310. Price, W. H. 1956. "The Isolation of a New Virus Associated with Respiratory Clinical Disease in Humans." *Proceedings of the National Academy of Sciences of the USA* 42: 892–896.

311. Wat, D. 2004. "The Common Cold: A Review of the Literature." *European Journal of Internal Medicine* 15: 79–88.

312. Andrewes, C. H. 1965. *The Common Cold*. London, UK: Weidenfeld and Nicolson.

313. Tyrrell, D. A. J., and R. M. Chanock. 1963. "Rhinoviruses: A Description." *Science* 141: 152–153.

314. ICTV. 2024. "Family: Picornaviridae." ICTV. Accessed September 9, 2024. https://ictv.global/report/chapter/picornaviridae/picornaviridae/enterovirus.

315. Hendley, J. O. 1998. "Epidemiology, Pathogenesis, and Treatment of the Common Cold." *Seminars in Pediatric Infectious Diseases* 9: 50–55.

316. Miller, E. K., et al. 2007. "Rhinovirus-Associated Hospitalizations in Young Children." *Journal of Infectious Diseases* 195: 773–781.

317. Jackson, D. J., et al. 2008. "Wheezing Rhinovirus Illnesses in Early Life Predict Asthma Development in High-Risk Children." *American Journal of Respiratory and Critical Care Medicine* 178: 667–672.

318. Lemanske Jr, R. F., et al. 2005. "Rhinovirus Illnesses During Infancy Predict Subsequent Childhood Wheezing." *Journal of Allergy and Clinical Immunology* 116: 571–577.

319. Ibid.

320. Michi, A. N., et al. 2020. "Rhinovirus-Induced Modulation of Epithelial Phenotype: Role in Asthma." *Viruses* 12: 1328.

321. Liu, L., et al. 2017. "Association Between Rhinovirus Wheezing Illness and the Development of Childhood Asthma: A Meta-Analysis." *BMJ Open* 7: e013034.

322. Skloot, R. 2010. *The Immortal Life of Henrietta Lacks*. New York: Random House.

323. NIAID. 2001. *Oral History Interview Project Interview #1 with Dr. Robert Chanock Conducted on January 11, 2001, by Peggy Dillon*. Bethesda, USA: NIAID.

324. Ibid.

325. Ibid.

326. Allen, A. 2007. *Vaccine*. New York: WW Norton & Company, Inc.

327. NIAID. 2001. *Oral History Interview Project Interview #1 with Dr. Robert Chanock Conducted on January 11, 2001, by Peggy Dillon*. Bethesda, USA: NIAID.

328. Chanock, R. M. 1956. "Association of a New Type of Cytopathogenic Myxovirus with Infantile Croup." *Journal of Experimental Medicine* 104: 555–576.

329. Wat, D. 2004. "The Common Cold: A Review of the Literature." *European Journal of Internal Medicine* 15: 79–88.

330. Kenmoe, S., et al. 2020. "Systematic Review and Meta-Analysis of the Prevalence of Common Respiratory Viruses in Children < 2 Years with Bronchiolitis in the Pre-COVID-19 Pandemic Era." *PLoS One* 15: e0242302.

331. Karron, R. A., and P. L. Collins. 2013. "Parainfluenza Viruses." In *Fields Virology*, edited by D. Knipe and P. Howley, 996–1023. Philadelphia, USA: Wolters Kluwer Health/Lippincott Williams & Wilkins.

332. Reed, G., et al. 1997. "Epidemiology and Clinical Impact of Parainfluenza Virus Infections in Otherwise Healthy Infants and Young Children < 5 Years Old." *Journal of Infectious Diseases* 175: 807–813.

333. Henrickson, K. J. 2003. "Parainfluenza Viruses." *Clinical Microbiology Reviews* 16: 242–264.

334. Morris, J. A., et al. 1956. "Recovery of Cytopathogenic Agent from Chimpanzees with Coryza." *Proceedings of the Society for Experimental Biology and Medicine* 92: 544–549.

335. Ibid.

336. Chanock, R., et al. 1957. "Recovery from Infants with Respiratory Illness of a Virus Related to Chimpanzee Coryza Agent (CCA). I. Isolation, Properties and Characterization." *American Journal of Hygiene* 66: 281–290.

337. Chanock, R., and L. Finberg. 1957. "Recovery from Infants with Respiratory Illness of a Virus Related to Chimpanzee Coryza Agent (CCA). II. Epidemiologic Aspects of Infection in Infants and Young Children." *American Journal of Hygiene* 66: 291–300.

338. Collins, P. L., and R. A. Karron. 2013. "Respiratory Syncytial Virus and Metapneumovirus." In *Fields Virology*, edited by D. Knipe and P. Howley, 1086–1123. Philadelphia, USA: Wolters Kluwer Health/Lippincott Williams & Wilkins.

339. Branche, A. R., and A. R. Falsey. 2016. "Parainfluenza Virus Infection." *Seminars in Respiratory and Critical Care Medicine* 37: 538–554.

340. NIAID. 2001. *Oral History Interview Project Interview #1 with Dr. Robert Chanock Conducted on January 11, 2001, by Peggy Dillon*. Bethesda, USA: NIAID.

341. Ibid.

342. Rima, B., et al. 2017. "ICTV Virus Taxonomy Profile: Pneumoviridae." *Journal of General Virology* 98: 2912–2913.

343. Wang, X., et al. 2024. "Global Disease Burden of and Risk Factors for Acute Lower Respiratory Infections Caused by Respiratory Syncytial Virus in Preterm Infants and Young Children in 2019: A Systematic Review and Meta-Analysis of Aggregated and Individual Participant Data." *Lancet* 403: 1241–1253.

344. Li, Y., et al. 2022. "Global, Regional, and National Disease Burden Estimates of Acute Lower Respiratory Infections Due to Respiratory Syncytial Virus in Children Younger Than 5 Years in 2019: A Systematic Analysis." *Lancet* 399: 2047–2064.

345. Ibid.

346. Collins, P. L., and R. A. Karron. 2013. "Respiratory Syncytial Virus and Metapneumovirus." In *Fields Virology*, edited by D. Knipe and P. Howley, 1086–1123. Philadelphia, USA: Wolters Kluwer Health/Lippincott Williams & Wilkins.

347. Allen, A. 2007. *Vaccine*. New York: WW Norton & Company, Inc.

348. Beecher, H. K. 1966. "Ethics and Clinical Research." *New England Journal of Medicine* 274: 1354–1360.

349. Bernard, D. 2019. "It Was Created as a Refuge for Needy Kids. Instead, They Were Raped and Drugged." *The Washington Post*, May 18, 2019.

350. Ibid.

351. NIAID. 2001. *Oral History Interview Project Interview #1 with Dr. Robert Chanock Conducted on January 11, 2001, by Peggy Dillon*. Bethesda, USA: NIAID.

352. Kapikian, A. Z., et al. 1969. "An Epidemiologic Study of Altered Clinical Reactivity to Respiratory Syncytial (RS) Virus Infection in Children Previously Vaccinated with an Inactivated RS Virus Vaccine." *American Journal of Epidemiology* 89: 405–421.

353. Wang, X., et al. 2024. "Global Disease Burden of and Risk Factors for Acute Lower Respiratory Infections Caused by Respiratory Syncytial Virus in Preterm Infants and Young Children in 2019: A Systematic Review and Meta-Analysis of Aggregated and Individual Participant Data." *Lancet* 403: 1241–1253.

354. Kim, H. W., et al. 1969. "Respiratory Syncytial Virus Disease in Infants Despite Prior Administration of Antigenic Inactivated Vaccine." *American Journal of Epidemiology* 89: 422–434.

355. Killikelly, A. M., et al. 2016. "Pre-Fusion F is Absent on the Surface of Formalin-Inactivated Respiratory Syncytial Virus." *Scientific Reports* 6: 34108.

356. Kampmann, B., et al. 2023. "Bivalent Prefusion F Vaccine in Pregnancy to Prevent RSV Illness in Infants." *New England Journal of Medicine* 388: 1451–1464.

357. CDC. 2023. "CDC Recommends New Vaccine to Help Protect Babies Against Severe Respiratory Syncytial Virus (RSV) Illness After Birth." CDC. Accessed September 19, 2024. https://www.cdc.gov/media/releases/2023/p0922-RSV-maternal-vaccine.html.

358. UKHSA. 2024. "Introduction of New NHS Vaccination Programmes Against Respiratory Syncytial Virus (RSV)." Gov.UK. Accessed September 19, 2024. https://www.gov.uk/government/publications/respiratory-syncytial-virus-rsv-vaccination-programmes-letter/introduction-of-new-nhs-vaccination-programmes-against-respiratory-syncytial-virus-rsv.

359. EMA. 2024. "Abrysvo." EMA. Accessed September 19, 2024. https://www.ema.europa.eu/en/medicines/human/EPAR/abrysvo.

360. Papi, A., et al. 2023. "Respiratory Syncytial Virus Prefusion F Protein Vaccine in Older Adults." *New England Journal of Medicine* 388: 595–608.

361. Walsh, E. E., et al. 2023. "Efficacy and Safety of a Bivalent RSV Prefusion F Vaccine in Older Adults." *New England Journal of Medicine* 388: 1465–1477.

362. Wilson, E., et al. 2023. "Efficacy and Safety of an mRNA-Based RSV PreF Vaccine in Older Adults." *New England Journal of Medicine* 389: 2233–2244.

363. Stafford, N. 2010. "Robert M Chanock." *British Medical Journal* 341: c6019.

364. Ibid.

365. Macmillan, M. H. 1960. "*PM Harold Macmillan – Wind of Change Speech at the Cape Town Parliament – 3 February 1960.*" Cape Town, South Africa. Accessed October 27, 2024. https://youtu.be/co7MiYfpOMw.

366. Kendall, E. J. C., et al. 1962. "Virus Isolations from Common Colds Occurring in a Residential School." *British Medical Journal* 2: 82–86.

367. Kerr, J. R., and D. Taylor-Robinson. 2007. "David Arthur John Tyrrell CBE: 19 June 1925–2 May 2005." *Biographical Memoirs of Fellows of the Royal Society* 53: 349–363.

368. Tyrrell, D. A. J. 1991. "Christopher Howard Andrewes, 7 June 1896–31 December 1987." *Biographical Memoirs of Fellows of the Royal Society* 37: 33–54.

369. Hoorn, B. 1964. "Respiratory Viruses in Model Experiments." *Acta Oto-Laryngologica* 57: S138–S144.

370. Tyrrell, D., and M. Fielder. 2002. *Cold Wars: The Fight Against the Common Cold.* Oxford: Oxford University Press.

371. Kerr, J. R., and D. Taylor-Robinson. 2007. "David Arthur John Tyrrell CBE: 19 June 1925–2 May 2005." *Biographical Memoirs of Fellows of the Royal Society* 53: 349–363.

372. Yaemsiri, S., et al. 2010. "Growth Rate of Human Fingernails and Toenails in Healthy American Young Adults." *Journal of the European Academy of Dermatology & Venereology* 24: 420–423.

373. Terrier, O., et al. 2009. "Parainfluenza Virus Type 5 (PIV-5) Morphology Revealed by Cryo-Electron Microscopy." *Virus Research* 142: 200–203.

374. Gil-Cantero, D., et al. 2024. "Cryo-EM of Human Rhinovirus Reveals Capsid-RNA Duplex Interactions That Provide Insights Into Virus Assembly and Genome Uncoating." *Communications Biology* 7: 1501.

375. Ruska, E. 1987. "Nobel Lecture. The Development of the Electron Microscope and of Electron Microscopy." *Bioscience Reports* 7: 607–629.

376. Von Kausche, G. A., and E. H. Ruska. 1939. "Die Sichtbarmachung von Pflanzlichem Virus im Übermikroskop." *Die Naturwissenschaften* 27: 292–299.

377. Auschwitz-Birkenau State Museum. 2016. "Bobrek." Auschwitz-Birkenau State Museum. Accessed August 19, 2025. https://www.auschwitz.org/en/history/auschwitz-sub-camps/bobrek/.

378. Best, J. M., et al. 1967. "Morphological Characteristics of Rubella Virus." *Lancet* 290: 237–239.

379. Tyrrell, D., and M. Fielder. 2002. *Cold Wars: The Fight Against the Common Cold.* Oxford: Oxford University Press.

380. Editorial. 1968. "Coronaviruses." *Nature* 220: 650.

381. Almeida, J. D., and D. A. J. Tyrrell. 1967. "The Morphology of Three Previously Uncharacterized Human Respiratory Viruses That Grow in Organ Culture." *Journal of General Virology* 1: 175–178; Tyrrell, D. A. J., and J. D. Almeida. 1967. "Direct Electron-Microscopy of Organ Culture for the Detection and Characterization of Viruses." *Archiv für die gesamte Virusforschung* 22: 417–425.

382. Hamre, D., and J. J. Procknow. 1966. "A New Virus Isolated from the Human Respiratory Tract." *Proceedings of the Society of Experimental Biology and Medicine* 121: 190–193.

383. McIntosh, K., et al. 1967. "Recovery in Tracheal Organ Cultures of Novel Viruses from Patients with Respiratory Disease." *Proceedings of the National Academy of Sciences of the USA* 57: 933–940.

384. Bradburne, A. F., et al. 1967. "Effects of a 'New' Human Respiratory Virus in Volunteers." *British Medical Journal* 3: 767–769.

385. Almeida, J. D., and D. A. J. Tyrrell. 1967. "The Morphology of Three Previously Uncharacterized Human Respiratory Viruses That Grow in Organ Culture." *Journal of General Virology* 1: 175–178.

386. Hamre, D., and M. Beem. 1972. "Virologic Studies of Acute Respiratory Disease in Young Adults. V. Coronavirus 229E Infections During Six Years of Surveillance." *American Journal of Epidemiology* 96: 94–106.

387. Hendley, J. O. 1998. "Epidemiology, Pathogenesis, and Treatment of the Common Cold." *Seminars in Pediatric Infectious Diseases* 9: 50–55.

388. van der Hoek, L. 2007. "Human Coronaviruses: What Do They Cause?" *Antiviral Therapy* 12: 651–658.

389. Callow, K. A., et al. 1990. "The Time Course of the Immune Response to Experimental Coronavirus Infection of Man." *Epidemiology and Infection* 105: 435–446.

390. Edridge, A. W. D., et al. 2020. "Seasonal Coronavirus Protective Immunity is Short-Lasting." *Nature Medicine* 26: 1691–1693.

391. Kerr, J. R., and D. Taylor-Robinson. 2007. "David Arthur John Tyrrell CBE: 19 June 1925–2 May 2005." *Biographical Memoirs of Fellows of the Royal Society* 53: 349–363.

392. Foot, G., and M. Hunsberger. 2020. "*The Virus Queen.*" San Antonio, USA: iHeart.

393. Kapikian, A. Z. 2005. "Remarks upon Acceptance of the 2005 Albert B. Sabin Gold Medal." In *2005 Albert B. Sabin Gold Medal*, edited by H. Shepherd, 14–28. Baltimore: Albert B. Sabin Vaccine Institute.

394. Kapikian, A. Z., et al. 1972. "Visualization by Immune Electron Microscopy of a 27-nm Particle Associated with Acute Infectious Nonbacterial Gastroenteritis." *Journal of Virology* 10: 1075–1081.

395. Almeida, J. 2008. "June Almeida (née Hart)." *British Medical Journal* 336: 1511.

396. Cherry, J. D., and P. Krogstad. 2004. "SARS: The First Pandemic of the 21st Century." *Pediatric Research* 56: 1–5.

397. WHO. 2015. "*Summary of Probable SARS Cases with Onset of Illness from 1 November 2002 to 31 July 2003.*" Geneva: WHO. Accessed October 27, 2024. https://www.who.int/publications/m/item/summary-of-probable-sars-cases-with-onset-of-illness-from-1-november-2002-to-31-july-2003.

398. Rosling, L., and M. Rosling. 2003. "Pneumonia Causes Panic in Guangdong Province." *British Medical Journal* 326: 416.

399. Tan, M., et al. 2024. "Aerosol Transmission of Norovirus." *Viruses* 16: 151.

400. Cherry, J. D., and P. Krogstad. 2004. "SARS: The First Pandemic of the 21st Century." *Pediatric Research* 56: 1–5.

401. WHO. 2015. "*Summary of Probable SARS Cases with Onset of Illness from 1 November 2002 to 31 July 2003.*" Geneva: WHO.

402. Heymann, D. L. 2004. "The International Response to the Outbreak of SARS in 2003." *Philosophical Transactions of the Royal Society of London B: Biological Sciences* 358: 1127–1129.

403. Cherry, J. D., and P. Krogstad. 2004. "SARS: The First Pandemic of the 21st Century." *Pediatric Research* 56: 1–5.

404. Ksiazek, T. G., et al. 2003. "A Novel Coronavirus Associated with Severe Acute Respiratory Syndrome." *New England Journal of Medicine* 348: 1953–1966.

405. Ibid.; Drosten, C., et al. 2003. "Identification of a Novel Coronavirus in Patients with Severe Acute Respiratory Syndrome." *New England Journal of Medicine* 348: 1967–1976.

406. Marra, M. A., et al. 2003. "The Genome Sequence of the SARS-Associated Coronavirus." *Science* 300: 1399–1404.

407. Fouchier, R. A. M., et al. 2004. "A Previously Undescribed Coronavirus Associated with Respiratory Disease in Humans." *Proceedings of the National Academy of Sciences of the USA* 101: 6212–6216.

408. van der Hoek, L., et al. 2004. "Identification of a New Human Coronavirus." *Nature Medicine* 10: 368–373.

409. Fouchier, R. A. 2024. Interview by author, October 30, 2024.

410. Ibid.

411. van der Hoek, L., et al. 2004. "Identification of a New Human Coronavirus." *Nature Medicine* 10: 368–373.

412. Fouchier, R. A. M., et al. 2004. "A Previously Undescribed Coronavirus Associated with Respiratory Disease in Humans." *Proceedings of the National Academy of Sciences of the USA* 101: 6212–6216.

413. Pyrc, K., et al. 2006. "Mosaic Structure of Human Coronavirus NL63, One Thousand Years of Evolution." *Journal of Molecular Biology* 364: 964–973.

414. Woo, P. C. Y., et al. 2005. "Characterization and Complete Genome Sequence of a Novel Coronavirus, Coronavirus HKU1, From Patients with Pneumonia." *Journal of Virology* 79: 884–895.

415. Woo, P. C. Y., et al. 2005. "Clinical and Molecular Epidemiological Features of Coronavirus HKU1-Associated Community-Acquired Pneumonia." *Journal of Infectious Diseases* 192: 1898–1907.

416. Wat, D. 2004. "The Common Cold: A Review of the Literature." *European Journal of Internal Medicine* 15: 79–88.

417. Poon, L. L. M., et al. 2004. "The Aetiology, Origins, and Diagnosis of Severe Acute Respiratory Syndrome." *Lancet Infectious Diseases* 4: 663–671.

418. Guan, Y., et al. 2003. "Isolation and Characterization of Viruses Related to the SARS Coronavirus from Animals in Southern China." *Science* 302: 276–278.

419. Cui, J., et al. 2019. "Origin and Evolution of Pathogenic Coronaviruses." *Nature Reviews Microbiology* 17: 181–192.

420. Irving, A. T., et al. 2021. "Lessons from the Host Defences of Bats, a Unique Viral Reservoir." *Nature* 589: 363–370.

421. Cohen, J. 2020. "Wuhan Coronavirus Hunter Shi Zhengli Speaks Out." *Science* 369: 487–488.

422. Li, H., et al. 2019. "Human-Animal Interactions and Bat Coronavirus Spillover Potential Among Rural Residents in Southern China." *Biosafety and Health* 1: 84–90.

423. Cui, J., et al. 2019. "Origin and Evolution of Pathogenic Coronaviruses." *Nature Reviews Microbiology* 17: 181–192.

424. Vijgen, L., et al. 2005. "Complete Genomic Sequence of Human Coronavirus OC43: Molecular Clock Analysis Suggests a Relatively Recent Zoonotic Coronavirus Transmission Event." *Journal of Virology* 79: 1595–1604.

425. Tao, Y., et al. 2017. "Surveillance of Bat Coronaviruses in Kenya Identifies Relatives of Human Coronaviruses NL63 and 229E and Their Recombination History." *Journal of Virology* 91: e01953–01916; Corman, V. M., et al. 2016. "Link of a Ubiquitous Human Coronavirus to Dromedary Camels." *Proceedings of the National Academy of Sciences of the USA* 113: 9864–9869.

426. Zaki, A. M., et al. 2012. "Isolation of a Novel Coronavirus from a Man with Pneumonia in Saudi Arabia." *New England Journal of Medicine* 367: 1814–1820.

427. WHO. 2024. "*MERS Situation Update, May 2024.*" Geneva: WHO. Accessed October 28, 2024. https://www.emro.who.int/health-topics/mers-cov/mers-outbreaks.html.

428. Hussein, I. 2014. "The Story of the First MERS Patient." *Nature Middle East* 2014: 134.

429. Müller, M. A., et al. 2014. "MERS Coronavirus Neutralizing Antibodies in Camels, Eastern Africa, 1983-1997." *Emerging Infectious Diseases* 20: 2093–2095.

430. WHO. 2025. "*MERS Situation Update, July 2025.*" Geneva: WHO. Accessed August 18, 2025. https://www.emro.who.int/health-topics/mers-cov/mers-outbreaks.html.

431. van den Hoogen, B. G., et al. 2001. "A Newly Discovered Human Pneumovirus Isolated from Young Children with Respiratory Tract Disease." *Nature Medicine* 7: 719–724.

432. Watson, J. D., and F. H. C. Crick. 1953. "Molecular Structure of Nucleic Acids; A Structure for Deoxyribose Nucleic Acid." *Nature* 171: 737–738.

433. Wilkins, M. H. F., et al. 1953. "Molecular Structure of Deoxypentose Nucleic Acids." *Nature* 171: 738–740.

434. Franklin, R. E., and R. G. Gosling. 1953. "Molecular Configuration in Sodium Thymonucleate." *Nature* 171: 740–741.

435. Jackson, Mick. 1987. *Life Story*. UK: BBC.

436. Watson, J. D. 1968. *The Double Helix: A Personal Account of the Discovery of the Structure of DNA*. London: Weidenfeld & Nicolson.

437. Creager, A. N. H., and G. J. Morgan. 2008. "After the Double Helix: Rosalind Franklin's Research on Tobacco Mosaic Virus." *Isis* 99: 239–272.

438. Crick, F. 1990. *What Mad Pursuit: A Personal View of Scientific Discovery*. London, UK: Penguin Books.

439. Cobb, M., and N. Comfort. 2023. "What Rosalind Franklin Truly Contributed to the Discovery of DNA's Structure." *Nature* 616: 657–660.

440. Hunt-Grubbe, C. 2007. "The Elementary DNA of Dr Watson." *Sunday Times*, October 14, 2007.

441. Editorial. 2007. "Watson Suspended Over Comments on Race." *Nature* 449: 960.

442. Basterfield, C., et al. 2020. "The Nobel Disease: When Intelligence Fails to Protect Against Irrationality." *Skeptical Inquirer* 44: 32–37.

443. Ambrogelly, A., et al. 2007. "Natural Expansion of the Genetic Code." *Nature Chemical Biology* 3: 29–35.

444. Creager, A. N. H., and G. J. Morgan. 2008. "After the Double Helix: Rosalind Franklin's Research on Tobacco Mosaic Virus." *Isis* 99: 239–272.

445. Mullis, K. B. 1990. "The Unusual Origin of the Polymerase Chain Reaction." *Scientific American* 262: 56–65.

446. Editorial. 1961. "Man of the Year: The Men on the Cover: US Scientists." *Time*, January 2, 1961.

447. Basterfield, C., et al. 2020. "The Nobel Disease: When Intelligence Fails to Protect Against Irrationality." *Skeptical Inquirer* 44: 32–37.

448. Mullis, K. 1968. "Cosmological Significance of Time Reversal." *Nature* 218: 663–664.

449. Mullis, K. 1998. *Dancing Naked in the Mind Field*. New York, USA: Random House.

450. Galton, F. 1904. "Eugenics; Its Definition, Scope and Aims." *Nature* 70: 82.

451. McDonald, C. 2019. "Intolerable Genius: Berkeley's Most Controversial Nobel Laureate." *California*, December 12, 2019.

452. Ibid.

453. Ibid.

454. Rabinow, P. 1996. *Making PCR : A Story of Biotechnology*. Chicago, USA: University of Chicago Press.

455. Ibid.

456. Ibid.

457. Kleppe, K., et al. 1971. "Studies on Polynucleotides. XCVI. Repair Replications of Short Synthetic DNA's as Catalyzed by DNA Polymerases." *Journal of Molecular Biology* 56: 341–361.

458. Rabinow, P. 1996. *Making PCR : A Story of Biotechnology*. Chicago, USA: University of Chicago Press.

459. Ibid.

460. Ibid.

461. Saiki, R. K., et al. 1985. "Enzymatic Amplification of B-Globin Genomic Sequences and Restriction Site Analysis for Diagnosis of Sickle Cell Anemia." *Science* 230: 1350–1354.

462. Mullis, K. B., and F. A. Faloona. 1987. "Specific Synthesis of DNA in Vitro via a Polymerase-Catalyzed Chain Reaction." *Methods in Enzymology* 155: 335–350.

463. Mullis, K., et al. 1986. "Specific Enzymatic Amplification of DNA in Vitro: The Polymerase Chain Reaction." *Cold Spring Harbor Symposia on Quantitative Biology* 51 Pt 1: 263–273.

464. Rabinow, P. 1996. *Making PCR : A Story of Biotechnology*. Chicago, USA: University of Chicago Press.

465. Ibid.

466. Bureau of Labor Statistics. 2025. "*CPI Inflation Calculator*." Washington DC: Bureau of Labor Statistics. Accessed November 19, 2025. https://data.bls.gov/cgi-bin/cpicalc.pl?cost1=1&year1=199106&year2=202506.

467. Xe Currency Converter. 2025. "*Historical Rate Tables*." Newmarket, Canada: Xe Corporation Inc. Accessed November 19, 2025. https://www.xe.com/currencytables/?from=USD&date=2025-07-01#table-section.

468. Bureau of Labor Statistics. 2025. "*CPI Inflation Calculator*." Washington DC: Bureau of Labor Statistics. Accessed November 19, 2025. https://data.bls.gov/cgi-bin/cpicalc.pl?cost1=1.00&year1=198606&year2=202506.

469. Xe Currency Converter. 2025. *"Historical Rate Tables."* Newmarket, Canada: Xe Corporation Inc. Accessed November 19, 2025. https://www.xe.com/currencytables/?from=USD&date=2025-07-01#table-section.

470. Wade, N. 1998. "Scientist at Work/Kary Mullis; After the 'Eureka,' a Nobelist Drops Out." *New York Times*, September 15, 1998.

471. Yoffe, E. 1994. "Is Kary Mullis God? (Or Just the Big Kahuna?)." *Esquire*, July 1, 1994.

472. Fieldhouse, R. 2024. "The Dark Side of the Maverick Genius Behind a Medical Revolution." *Australian Doctor*, February 17, 2024.

473. Yoffe, E. 1994. "Is Kary Mullis God? (Or Just the Big Kahuna?)." *Esquire*, July 1, 1994.

474. Rabinow, P. 1996. *Making PCR : A Story of Biotechnology*. Chicago, USA: University of Chicago Press.

475. Kwok, S., et al. 1987. "Identification of Human Immunodeficiency Virus Sequences by Using in Vitro Enzymatic Amplification and Oligomer Cleavage Detection." *Journal of Virology* 61: 1690–1694.

476. Yoffe, E. 1994. "Is Kary Mullis God? (Or Just the Big Kahuna?)." *Esquire*, July 1, 1994.

477. Martin, J. F. 1994. "Bad Example." *Nature* 371: 97.

478. Fieldhouse, R. 2024. "The Dark Side of the Maverick Genius Behind a Medical Revolution." *Australian Doctor*, February 17, 2024.

479. World Bank. 2025. *"Prevalence of HIV, Total (% of Population Ages 15-49) – South Africa."* Washington DC: World Bank Group. Accessed August 22, 2025. https://data.worldbank.org/indicator/SH.DYN.AIDS.ZS?locations=ZA.

480. Fieldhouse, R. 2024. "The Dark Side of the Maverick Genius Behind a Medical Revolution." *Australian Doctor*, February 17, 2024.

481. Chigwedere, P., et al. 2008. "Estimating the Lost Benefits of Antiretroviral Drug Use in South Africa." *Journal of Acquired Immune Deficiency Syndrome* 49: 410–415.

482. McDonald, C. 2019. "Intolerable Genius: Berkeley's Most Controversial Nobel Laureate." *California*, December 12, 2019.

483. Mullis, K. 1998. *Dancing Naked in the Mind Field*. New York, USA: Random House.

484. Basterfield, C., et al. 2020. "The Nobel Disease: When Intelligence Fails to Protect Against Irrationality." *Skeptical Inquirer* 44: 32–37.

485. Rima, B., et al. 2017. "ICTV Virus Taxonomy Profile: Pneumoviridae." *Journal of General Virology* 98: 2912–2913.

486. van den Hoogen, B. G., et al. 2001. "A Newly Discovered Human Pneumovirus Isolated from Young Children with Respiratory Tract Disease." *Nature Medicine* 7: 719–724.

487. Hermos, C. R., et al. 2010. "Human Metapneumovirus." *Clinics in Laboratory Medicine* 30: 131–148.

488. Moriyama, M., et al. 2020. "Seasonality of Respiratory Viral Infections." *Annual Review of Virology* 7: 83–101.

489. Allander, T., et al. 2005. "Cloning of a Human Parvovirus by Molecular Screening of Respiratory Tract Samples." *Proceedings of the National Academy of Sciences of the USA* 102: 12891–12896.

490. Trapani, S., et al. 2023. "Human Bocavirus in Childhood: A True Respiratory Pathogen or a 'Passenger' Virus? A Comprehensive Review." *Microorganisms* 11: 1243.

491. Colazo Salbetti, M. B., et al. 2023. "Human Bocavirus Respiratory Infection: Tracing the Path from Viral Replication and Virus-Cell Interactions to Diagnostic Methods." *Reviews in Medical Virology* 33: e2482.

492. Da Silva, S. J. R., et al. 2023. "Recent Insights Into SARS-CoV-2 Omicron Variant." *Reviews in Medical Virology* 33: e2373.

493. Akhtar, A. 2025. Email to author, August 29, 2025.

494. Ibid.

495. Ibid.

496. Taubenberger, J. K., et al. 2007. "Discovery and Characterization of the 1918 Pandemic Influenza Virus in Historical Context." *Antiviral Therapy* 12: 581–591.

497. Zhou, P., et al. 2020. "A Pneumonia Outbreak Associated with a New Coronavirus of Probable Bat Origin." *Nature* 579: 270–273.

498. O'Driscoll, M., et al. 2020. "Age-Specific Mortality and Immunity Patterns of SARS-CoV-2." *Nature* 590: 140–145.

499. Taubenberger, J. K., and D. M. Morens. 2006. "1918 Influenza: The Mother of All Pandemics." *Emerging Infectious Diseases* 12: 15–22.

500. Taubenberger, J. K., et al. 2007. "Discovery and Characterization of the 1918 Pandemic Influenza Virus in Historical Context." *Antiviral Therapy* 12: 581–591.

501. Sridhar, D. L. 2022. *Preventable: How a Pandemic Changed the World & How to Stop the Next One*. London, UK: Penguin Random House UK.

502. Covid-19 Excess Mortality Collaborators. 2022. "Estimating Excess Mortality Due to the COVID-19 Pandemic: A Systematic Analysis of COVID-19-Related Mortality, 2020-21." *Lancet* 399: 1513–1536.

503. Callow, K. A., et al. 1990. "The Time Course of the Immune Response to Experimental Coronavirus Infection of Man." *Epidemiology and Infection* 105: 435–446.

504. Edridge, A. W. D., et al. 2020. "Seasonal Coronavirus Protective Immunity is Short-Lasting." *Nature Medicine* 26: 1691–1693.

505. Graham, M. S., et al. 2021. "Changes in Symptomatology, Reinfection, and Transmissibility Associated with the SARS-CoV-2 Variant B.1.1.7: An Ecological Study." *Lancet Public Health* 6: e335–e345.

506. Cameroni, E., et al. 2022. "Broadly Neutralizing Antibodies Overcome SARS-CoV-2 Omicron Antigenic Shift." *Nature* 602: 664–670.

507. Wolf, J. M., et al. 2023. "Molecular Evolution of SARS-CoV-2 from December 2019 to August 2022." *Journal of Medical Virology* 95: e28366.

508. Cameroni, E., et al. 2022. "Broadly Neutralizing Antibodies Overcome SARS-CoV-2 Omicron Antigenic Shift." *Nature* 602: 664–670.

509. Da Silva, S. J. R., et al. 2023. "Recent Insights Into SARS-CoV-2 Omicron Variant." *Reviews in Medical Virology* 33: e2373.

510. Gili, R., and R. Burioni. 2023. "SARS-CoV-2 Before and After Omicron: Two Different Viruses and Two Different Diseases?" *Journal of Translational Medicine* 21: 251.

511. Gordon, N. 2022. "How a 'Forgotten,' 110-Day Lockdown Sparked China's Nationwide COVID Protests." *Fortune*, November 28, 2022.

512. Pollard, M. Q. 2022. "Chinese Cities Ease Curbs, Full Zero-COVID Exit Seen Some Way Off." *Reuters*. Accessed February 25, 2025. https://www.reuters.com/world/china/chinese-cities-announce-further-easing-covid-curbs-2022-12-04/.

513. Fu, Y., et al. 2025. "Effectiveness and Coverage of COVID-19 Vaccination Among the Infection-Naive Population: A Community-Based Retrospective Cohort Study in China." *Vaccine* 50: 126836.

514. Chen, X., et al. 2022. "Estimation of Disease Burden and Clinical Severity of COVID-19 Caused by Omicron BA.2 in Shanghai, February-June 2022." *Emerging Microbes & Infections* 11: 2800–2807.

515. Deng, Y., et al. 2023. "The Risks of Death and Hospitalizations Associated with SARS-CoV-2 Omicron Declined After Lifting Testing and Quarantining Measures." *Journal of Infection* 86: e123–e125.

516. O'Driscoll, M., et al. 2020. "Age-Specific Mortality and Immunity Patterns of SARS-CoV-2." *Nature* 590: 140–145.

517. Du, Z., et al. 2023. "Estimate of COVID-19 Deaths, China, December 2022-February 2023." *Emerging Infectious Diseases* 29: 2121–2124.

518. Kistler, K. E., and T. Bedford. 2023. "An Atlas of Continuous Adaptive Evolution in Endemic Human Viruses." *Cell Host & Microbe* 31: 1898–1909 e1893.

519. UKHSA. 2025. "*UKHSA Data Dashboard: COVID-19.*" London: UKHSA. Accessed August 27, 2025. https://ukhsa-dashboard.data.gov.uk/respiratory-viruses/covid-19#deaths.

520. Da Silva, S. J. R., et al. 2023. "Recent Insights Into SARS-CoV-2 Omicron Variant." *Reviews in Medical Virology* 33: e2373.

521. Chemaitelly, H., et al. 2025. "Differential Protection Against SARS-CoV-2 Reinfection Pre- and Post-Omicron." *Nature*.

522. Sauerwein, K. 2023. "Real-World Reflections: WashU Medicine Nephrologist Has Risen to Prominence, Uncovering Society's Biggest Health Issues." *Outlook* (WashU Medicine), Spring 2023.

523. Sprayregen, M. 2020. "This Support Group Is Helping Launch A Covid-19 Patient Advocacy Movement." *Forbes*, November 13, 2020.

524. Lowenstein, F. 2020. "We Need to Talk About What Coronavirus Recoveries Look Like." *New York Times*, April 13, 2020.

525. Ibid.

526. Zimmer, B. 2021. "'Long-Hauler': When Covid-19's Symptoms Last and Last." *Wall Street Journal*, January 1, 2021.

527. Porter, K. 2020. "*Oh That's Funny... It Absolutely Came From Amy. She Named the Group That Because of Her Favorite Trucker Hat She Wore When She Got Tested.*" Twitter. Accessed May 17, 2022. https://twitter.com/katemeredithp/status/1277316840453267456.

528. Al-Aly, Z., et al. 2021. "High-Dimensional Characterization of Post-Acute Sequelae of COVID-19." *Nature* 594: 259–264.

529. Dixon, L. 2024. "*What Is Long-COVID? Dr. Ziyad Al-Aly Explains the Symptoms and Science of Long-COVID.*" Canada: Friendly Pharmacy 5.

530. Ibid.

531. Akhtar, A. 2025. Email to author, August 29, 2025.

532. Hakki, S., et al. 2022. "Onset and Window of SARS-CoV-2 Infectiousness and Temporal Correlation with Symptom Onset: A Prospective, Longitudinal, Community Cohort Study." *Lancet Respiratory Medicine* 10: 1061–1073.

533. Peluso, M. J., et al. 2021. "Markers of Immune Activation and Inflammation in Individuals With Postacute Sequelae of Severe Acute Respiratory Syndrome Coronavirus 2 Infection." *Journal of Infectious Diseases* 224: 1839–1848.

534. Pretorius, E., et al. 2022. "Prevalence of Symptoms, Comorbidities, Fibrin Amyloid Microclots and Platelet Pathology in Individuals with Long COVID/Post-Acute Sequelae of COVID-19 (PASC)." *Cardiovascular Diabetology* 21: 148.

535. Turner, S., et al. 2023. "Long COVID: Pathophysiological Factors and Abnormalities of Coagulation." *Trends in Endocrinology & Metabolism*.

536. Noonong, K., et al. 2023. "Mitochondrial Oxidative Stress, Mitochondrial ROS Storms in Long COVID Pathogenesis." *Frontiers of Immunology* 14: 1275001.

537. Akhtar, A. 2025. Email to author, August 29, 2025.

538. Greene, C., et al. 2024. "Blood-Brain Barrier Disruption and Sustained Systemic Inflammation in Individuals with Long COVID-Associated Cognitive Impairment." *Nature Neuroscience* 27: 421–432.

539. Davis, H. E., et al. 2023. "Long COVID: Major Findings, Mechanisms and Recommendations." *Nature Reviews Microbiology* 21: 133–146.

540. Weinstock, L. B., et al. 2021. "Mast Cell Activation Symptoms Are Prevalent in Long-COVID." *International Journal of Infectious Diseases* 112: 217–226.

541. Sender, R., et al. 2016. "Revised Estimates for the Number of Human and Bacteria Cells in the Body." *PLoS Biology* 14: e1002533.

542. Ailioaie, L. M., et al. 2023. "Gut Microbiota and Mitochondria: Health and Pathophysiological Aspects of Long COVID." *International Journal of Molecular Sciences* 24: 17198.

543. Lau, R. I., et al. 2024. "A Synbiotic Preparation (SIM01) for Post-Acute COVID-19 Syndrome in Hong Kong (RECOVERY): A Randomised, Double-Blind, Placebo-Controlled Trial." *Lancet Infectious Diseases* 24: 256–265.

544. Proal, A. D., et al. 2023. "SARS-CoV-2 Reservoir in Post-Acute Sequelae of COVID-19 (PASC)." *Nature Immunology* 24: 1616–1627.

545. Proal, A. D., et al. 2025. "Targeting the SARS-CoV-2 Reservoir in Long COVID." *Lancet Infectious Diseases* 25: e294–e306.

546. Friendly Pharmacy 5. 2024. "What Is Long-COVID? Dr. Ziyad Al-Aly Explains the Symptoms and Science of Long-COVID." YouTube. https://www.youtube.com/watch?v=dpbeuwHKQWo.

547. WHO. 2023. "*Statement on the Fifteenth Meeting of the IHR (2005) Emergency Committee on the COVID-19 Pandemic.*" Geneva: WHO. Accessed August 28, 2025. https://www.who.int/news/item/05-05-2023-statement-on-the-fifteenth-meeting-of-the-international-health-regulations-(2005)-emergency-committee-regarding-the-coronavirus-disease-(covid-19)-pandemic.

548. UKHSA. 2025. "*UKHSA Data Dashboard: COVID-19.*" London: UKHSA. Accessed August 27, 2025. https://ukhsa-dashboard.data.gov.uk/respiratory-viruses/covid-19#deaths.

549. UKHSA. 2023. "*COVID-19 Infection Survey Participants Thanked for 'Huge Contribution' to Pandemic Response.*" London: UKHSA. Accessed February 25, 2025. https://www.gov.uk/government/news/covid-19-infection-survey-participants-thanked-for-huge-contribution-to-pandemic-response.

550. Xie, Y., et al. 2024. "Postacute Sequelae of SARS-CoV-2 Infection in the Pre-Delta, Delta, and Omicron Eras." *New England Journal of Medicine* 391: 515–525.

551. Bobrovitz, N., et al. 2023. "Protective Effectiveness of Previous SARS-CoV-2 Infection and Hybrid Immunity Against the Omicron Variant and Severe Disease: A Systematic Review and Meta-Regression." *Lancet Infectious Diseases* 23: 556–567.

552. Català, M., et al. 2024. "The Effectiveness of COVID-19 Vaccines to Prevent Long COVID Symptoms: Staggered Cohort Study of Data from the UK, Spain, and Estonia." *Lancet Respiratory Medicine* 12: 225–236.

553. Marra, A. R., et al. 2023. "The Effectiveness of COVID-19 Vaccine in the Prevention of Post-COVID Conditions: A Systematic Literature Review and Meta-Analysis of the Latest Research." *Antimicrobial Stewardship & Healthcare Epidemiology* 3: e168.

554. Al-Aly, Z., et al. 2024. "Long COVID Science, Research and Policy." *Nature Medicine* 30: 2148–2164.

555. Greenhalgh, T., et al. 2022. "How Covid-19 Spreads: Narratives, Counter Narratives, and Social Dramas." *British Medical Journal* 378: e069940.

556. Al-Aly, Z., et al. 2024. "Long COVID Science, Research and Policy." *Nature Medicine* 30: 2148–2164.

557. ONS. 2023. "*Prevalence of Ongoing Symptoms Following Coronavirus (COVID-19) Infection in the UK: 30 March 2023.*" London: Office for National Statistics. Accessed February 25, 2025. https://www.ons.gov.uk/peoplepopulationandcommunity/healthandsocialcare/conditionsanddiseases/bulletins/prevalenceofongoingsymptomsfollowingcoronaviruscovid19infectionintheuk/30march2023.

558. Liu, C., et al. 2025. "Association of COVID Vaccinations and Treatments with Long COVID Beyond 6 Months: A Case-Control Study on the Adult Population in a Large Integrated Healthcare System in the United States from 2020 to 2023." *Preventive Medicine Reports* 57: 103188.

559. MacIntyre, C. R. 2025. Interview by author, July 14, 2025.

560. Griffin, S. 2025. Interview by author, July 28, 2025.

561. Akhtar, A. 2025. Email to author, February 21, 2025.

562. Klein, R. M. 2022. "Calvin Coolidge's 'Rose Fever': Immunotherapy and Chlorine Gas Treatment." *Annals of Allergy, Asthma & Immunology* 129: 125–126.

563. Hunter, P. 2007. "A Question of Faith. Exploiting the Placebo Effect Depends on Both the Susceptibility of the Patient to Suggestion and the Ability of the Doctor to Instill Trust." *EMBO Reports* 8: 125–128.

564. MRC. 1944. "Clinical Trial of Patulin in the Common Cold. Report of the Patulin Clinical Trials Committee, Medical Research Council." *Lancet* 244: 373–375.

565. Seida, J. K., et al. 2011. "North American (Panax Quinquefolius) and Asian Ginseng (Panax Ginseng) Preparations for Prevention of the Common Cold in Healthy Adults: A Systematic Review." *Evidence-Based Complementary and Alternative Medicine* 2011: 282151.

566. Karsch-Völk, M., et al. 2014. "Echinacea for Preventing and Treating the Common Cold." *Cochrane Database of Systematic Reviews* 2014: CD000530.

567. Dunitz, J. D. 1996. "Linus Carl Pauling: 28 February 1901–19 August 1994." *Biographical Memoirs of Fellows of the Royal Society* 42: 317–338.

568. Pauling, L. 1970. *Vitamin C and the Common Cold*. New York, USA: WH Freeman & Co.

569. Basterfield, C., et al. 2020. "The Nobel Disease: When Intelligence Fails to Protect Against Irrationality." *Skeptical Inquirer* 44: 32–37.

570. Pauling, L. 1971. "The Significance of the Evidence About Ascorbic Acid and the Common Cold." *Proceedings of the National Academy of Sciences of the USA* 68: 2678–2681.

571. Cameron, E., and L. Pauling. 1976. "Supplemental Ascorbate in the Supportive Treatment of Cancer: Prolongation of Survival Times in Terminal Human Cancer." *Proceedings of the National Academy of Sciences of the USA* 73: 3685–3689.

572. Dunitz, J. D. 1996. "Linus Carl Pauling: 28 February 1901–19 August 1994." *Biographical Memoirs of Fellows of the Royal Society* 42: 317–338.

573. Hemilä, H., and E. Chalker. 2013. "Vitamin C for Preventing and Treating the Common Cold." *Cochrane Database of Systematic Reviews* 2013: CD000980.

574. Hemilä, H. 2017. "Vitamin C and Infections." *Nutrients* 9: 339.

575. Cerullo, G., et al. 2020. "The Long History of Vitamin C: From Prevention of the Common Cold to Potential Aid in the Treatment of COVID-19." *Frontiers in Immunology* 11: 574029.

576. Carr, A. C., et al. 2013. "Human Skeletal Muscle Ascorbate is Highly Responsive to Changes in Vitamin C Intake and Plasma Concentrations." *American Journal of Clinical Nutrition* 97: 800–807.

577. Beard, J. A., et al. 2011. "Vitamin D and the Anti-Viral State." *Journal of Clinical Virology* 50: 194–200.

578. Sutherland, J. P., et al. 2021. "Differences and Determinants of Vitamin D Deficiency Among UK Biobank Participants: A Cross-Ethnic and Socioeconomic Study." *Clinical Nutrition* 40: 3436–3447.

579. Martineau, A. R., et al. 2017. "Vitamin D Supplementation to Prevent Acute Respiratory Tract Infections: Systematic Review and Meta-Analysis of Individual Participant Data." *British Medical Journal* 356: i6583.

580. Read, S. A., et al. 2019. "The Role of Zinc in Antiviral Immunity." *Advances in Nutrition* 10: 696–710.

581. Singh, M., and R. R. Das. 2013. "Zinc for the Common Cold." *Cochrane Database of Systematic Reviews*: CD001364.

582. Ibid.

583. Ibid.

584. Sender, R., et al. 2016. "Revised Estimates for the Number of Human and Bacteria Cells in the Body." *PLoS Biology* 14: e1002533.

585. Lehtinen, M. J., et al. 2018. "Nasal Microbiota Clusters Associate with Inflammatory Response, Viral Load, and Symptom Severity in Experimental Rhinovirus Challenge." *Scientific Reports* 8: 11411.

586. Zhao, Y., et al. 2022. "Probiotics for Preventing Acute Upper Respiratory Tract Infections." *Cochrane Database of Systematic Reviews* 8: CD006895.

587. Eccles, R. 2009. "Over the Counter Medicines for Colds." In *Common Cold*, edited by R. Eccles and O. Weber, 249–273. Basel, Switzerland: Birkhäuser.

588. Eccles, R. 2006. "Efficacy and Safety of Over-The-Counter Analgesics in the Treatment of Common Cold and Flu." *Journal of Clinical Pharmacy and Therapeutics* 31: 309–319.

589. Little, P., et al. 2013. "Ibuprofen, Paracetamol, and Steam for Patients with Respiratory Tract Infections in Primary Care: Pragmatic Randomised Factorial Trial." *British Medical Journal* 347: f6041.

590. Eccles, R. 2005. "Understanding the Symptoms of the Common Cold and Influenza." *Lancet Infectious Diseases* 5: 718–725.

591. De Sutter, A. I., et al. 2015. "Antihistamines for the Common Cold." *Cochrane Database of Systematic Reviews* 2015: CD009345.

592. Deckx, L., et al. 2016. "Nasal Decongestants in Monotherapy for the Common Cold." *Cochrane Database of Systematic Reviews* 10: CD009612.

593. Kollar, C., et al. 2007. "Meta-Analysis of the Efficacy of a Single Dose of Phenylephrine 10 mg Compared with Placebo in Adults with Acute Nasal Congestion Due to the Common Cold." *Clinical Therapeutics* 29: 1057–1070.

594. Eccles, R., et al. 2005. "Efficacy and Safety of Single and Multiple Doses of Pseudoephedrine in the Treatment of Nasal Congestion Associated with Common Cold." *American Journal of Rhinology* 19: 25–31.

595. Eccles, R., et al. 2010. "Effects of Intranasal Xylometazoline, Alone or in Combination with Ipratropium, in Patients with Common Cold." *Current Medical Research and Opinion* 26: 889–899.

596. AlBalawi, Z. H., et al. 2013. "Intranasal Ipratropium Bromide for the Common Cold." *Cochrane Database of Systematic Reviews* 2013: CD008231.

597. Eccles, R., et al. 2010. "Effects of Intranasal Xylometazoline, Alone or in Combination with Ipratropium, in Patients with Common Cold." *Current Medical Research and Opinion* 26: 889–899.

598. Eccles, R. 2009. "Over the Counter Medicines for Colds." In *Common Cold*, edited by R. Eccles and O. Weber, 249–273. Basel, Switzerland: Birkhäuser.

599. Eccles, R. 2025. Interview by author, March 13, 2025.

600. Paul, I. M., et al. 2007. "Effect of Honey, Dextromethorphan, and No Treatment on Nocturnal Cough and Sleep Quality for Coughing Children and Their Parents." *Archives of Pediatrics & Adolescent Medicine* 161: 1140–1146.

601. Eccles, R. 2006. "Mechanisms of the Placebo Effect of Sweet Cough Syrups." *Respiratory Physiology & Neurobiology* 152: 340–348.

602. Ibid.

603. Eccles, R. 2025. Interview by author, March 13, 2025.

604. Munger, R. S. 1949. "Guaiacum, the Holy Wood From the New World." *Journal of the History of Medicine and Allied Sciences* 4: 196–229.

605. Albrecht, H. H., et al. 2017. "Role of Guaifenesin in the Management of Chronic Bronchitis and Upper Respiratory Tract Infections." *Multidisciplinary Respiratory Medicine* 12: 31.

606. Bennett, W. D., et al. 2015. "Effect of a Single 1200 Mg Dose of Mucinex(R) on Mucociliary and Cough Clearance During an Acute Respiratory Tract Infection." *Respiratory Medicine* 109: 1476–1483.

607. Albrecht, H., et al. 2012. "Patient-Reported Outcomes to Assess the Efficacy of Extended-Release Guaifenesin for the Treatment of Acute Respiratory Tract Infection Symptoms." *Respiratory Research* 13: 118.

608. Singh, M., et al. 2017. "Heated, Humidified Air for the Common Cold." *Cochrane Database Systematic Reviews* 8: CD001728.

609. Eccles, R. 2009. "Over the Counter Medicines for Colds." In *Common Cold*, edited by R. Eccles and O. Weber, 249–273. Basel, Switzerland: Birkhäuser.

610. Satomura, K., et al. 2005. "Prevention of Upper Respiratory Tract Infections by Gargling: A Randomized Trial." *American Journal of Preventive Medicine* 29: 302–307.

611. Ramalingam, S., et al. 2019. "A Pilot, Open Labelled, Randomised Controlled Trial of Hypertonic Saline Nasal Irrigation and Gargling for the Common Cold." *Scientific Reports* 9: 1015.

612. Ahmad, L. 2021. "Impact of Gargling on Respiratory Infections." *All Life* 14: 147–158.

613. Ide, K., et al. 2016. "Effect of Gargling with Tea and Ingredients of Tea on the Prevention of Influenza Infection: A Meta-Analysis." *BMC Public Health* 16: 396.

614. Furushima, D., et al. 2018. "Effect of Tea Catechins on Influenza Infection and the Common Cold with a Focus on Epidemiological/Clinical Studies." *Molecules* 23: 1795.

615. Chong, L. Y., et al. 2016. "Saline Irrigation for Chronic Rhinosinusitis." *Cochrane Database Systematic Reviews* 4: CD011995.

616. King, D., et al. 2015. "Saline Nasal Irrigation for Acute Upper Respiratory Tract Infections." *Cochrane Database Systematic Reviews* 2015: CD006821.

617. Chong, L. Y., et al. 2016. "Saline Irrigation for Chronic Rhinosinusitis." *Cochrane Database Systematic Reviews* 4: CD011995.

618. Hemilä, H., and E. Chalker. 2021. "Carrageenan Nasal Spray May Double the Rate of Recovery from Coronavirus and Influenza Virus Infections: Re-Analysis of Randomized Trial Data." *Pharmacology Research & Perspectives* 9: e00810.

619. Little, P., et al. 2024. "Nasal Sprays and Behavioural Interventions Compared with Usual Care for Acute Respiratory Illness in Primary Care: A Randomised, Controlled, Open-Label, Parallel-Group Trial." *Lancet Respiratory Medicine* 12: 619–632.

620. Ong, H.-H., et al. 2025. "Inhibitory Activity of Hydroxypropyl Methylcellulose on Rhinovirus and Influenza A Virus Infection of Human Nasal Epithelial Cells." *Viruses* 17: 376.

621. Foot, G. 2023. "*Cold Defence Nasal Sprays.*" *Sliced Bread*. London: BBC Radio 4. https://www.bbc.co.uk/programmes/m001gl6g.

622. Albrecht, H., et al. 2012. "Patient-Reported Outcomes to Assess the Efficacy of Extended-Release Guaifenesin for the Treatment of Acute Respiratory Tract Infection Symptoms." *Respiratory Research* 13: 118.

623. Eccles, R. 2007. "The Power of the Placebo." *Current Allergy and Asthma Reports* 7: 100–104.

624. Eccles, R. 2025. Interview by author, March 13, 2025.

625. Jessney, B. 2012. "Joseph Lister (1827-1912): A Pioneer of Antiseptic Surgery Remembered a Century After His Death." *Journal of Medical Biography* 20: 107–110.

626. Zajaczkowski, T. 2008. "Johann Anton von Mikulicz-Radecki (1850-1905)–A Pioneer of Gastroscopy and Modern Surgery: His Credit to Urology." *World Journal of Urology* 26: 75–86.

627. Schlich, T., and B. J. Strasser. 2022. "Making the Medical Mask: Surgery, Bacteriology, and the Control of Infection (1870s–1920s)." *Medical History* 66: 116–134.

628. Ibid.

629. Mikulicz, J. 1897. "Das Operieren in Sterilisierten Zwirnhandschuhen und mit Mundbinde." *Centralblatt für Chirurgie* 26: 713–717.

630. Bandou, R., et al. 2022. "Higher Viral Stability and Ethanol Resistance of Avian Influenza A(H5N1) Virus on Human Skin." *Emerging Infectious Diseases* 28: 639–649.

631. Watanabe, N., et al. 2023. "Evaluation of Environmental Stability and Disinfectant Effectiveness for Human Coronavirus OC43 on Human Skin Surface." *Microbiology Spectrum* 11: e0238122.

632. Aiello, A. E., et al. 2008. "Effect of Hand Hygiene on Infectious Disease Risk in the Community Setting: A Meta-Analysis." *American Journal of Public Health* 98: 1372–1381.

633. Greenhalgh, T., et al. 2022. "How Covid-19 Spreads: Narratives, Counter Narratives, and Social Dramas." *British Medical Journal* 378: e069940.

634. Hirose, R., et al. 2021. "Survival of Severe Acute Respiratory Syndrome Coronavirus 2 (SARS-CoV-2) and Influenza Virus on Human Skin: Importance of Hand Hygiene in Coronavirus Disease 2019 (COVID-19)." *Clinical Infectious Diseases* 73: e4329–e4335.

635. Heikkinen, T., and M. D. Järvinen. 2003. "The Common Cold." *Lancet* 361: 51–59.

636. Savolainen-Kopra, C., et al. 2012. "Single Treatment with Ethanol Hand Rub Is Ineffective Against Human Rhinovirus–Hand Washing with Soap and Water Removes the Virus Efficiently." *Journal of Medical Virology* 84: 543–547.

637. Wood, A., and D. Payne. 1998. "The Action of Three Antiseptics/Disinfectants Against Enveloped and Non-Enveloped Viruses." *Journal of Hospital Infection* 38: 283–295.

638. Zhang, R., et al. 2020. "Identifying Airborne Transmission as the Dominant Route for the Spread of COVID-19." *Proceedings of the National Academy of Sciences of the USA* 117: 14857–14863.

639. Schlich, T., and B. J. Strasser. 2022. "Making the Medical Mask: Surgery, Bacteriology, and the Control of Infection (1870s–1920s)." *Medical History* 66: 116–134.

640. Ibid.

641. Wu, X. 1987. "Dr Wu Lian-Teh, Renowned Epidemiologist and Pioneer of Modern Health Work in China." *Chinese Medical Journal* 100: 506–517.

642. Gamsa, M. 2006. "The Epidemic of Pneumonic Plague in Manchuria 1910-1911." *Past & Present* 190: 147–183.

643. Lynteris, C. 2018. "Plague Masks: The Visual Emergence of Anti-Epidemic Personal Protection Equipment." *Medical Anthropology* 37: 442–457.

644. Gainty, C., and J. Olszynko-Gryn. 2023. *"Mask Ambivalence in the 1918 Influenza Pandemic."* History Workshop. Accessed September 28, 2023. https://www.historyworkshop.org.uk/science-medicine-health/mask-ambivalence-in-the-1918-influenza-pandemic/.

645. Berger, K. 2020. *"The Mask Wars of the 1918 Flu Pandemic."* Crosscut. Accessed September 28, 2023. https://crosscut.com/2020/07/mask-wars-1918-flu-pandemic.

646. Schlich, T., and B. J. Strasser. 2022. "Making the Medical Mask: Surgery, Bacteriology, and the Control of Infection (1870s–1920s)." *Medical History* 66: 116–134.

647. Corbett, K. 2020. "True Story: A Former House Beautiful Editor Inspired the N95 Mask While Designing Bras." *House Beautiful*, May 28, 2020.

648. Adanur, S., and A. Jayswal. 2020. "Filtration Mechanisms and Manufacturing Methods of Facemasks: An Overview." *Journal of Industrial Textiles* 51: 3683S–3717S.

649. Wilson, M. 2020. "The Untold Origin Story of the N95 Mask." *Fast Company*, March 24, 2020.

650. Institute of Medicine. 2006. *Reusability of Facemasks During an Influenza Pandemic: Facing the Flu.* Washington DC: National Academy of Sciences.

651. Locke, L., et al. 2021. "Aerosol Transmission of Infectious Disease and the Efficacy of Personal Protective Equipment (PPE): A Systematic Review." *Journal of Occupational and Environmental Medicine* 63: e783–e791.

652. Clase, C. M., et al. 2020. "Forgotten Technology in the COVID-19 Pandemic: Filtration Properties of Cloth and Cloth Masks-A Narrative Review." *Mayo Clinic Proceedings* 95: 2204–2224.

653. Du, W., et al. 2021. "Microstructure Analysis and Image-Based Modelling of Facemasks for COVID-19 Virus Protection." *Communications Materials* 2: 69.

654. Wu, J., et al. 2004. "Risk Factors for SARS Among Persons Without Known Contact with SARS Patients, Beijing, China." *Emerging Infectious Diseases* 10: 210–216.

655. Chen, H., et al. 2022. "Can Disposable Masks Be Worn More Than Once?" *Ecotoxicology and Environmental Safety* 242: 113908.

656. Checchi, V., et al. 2021. "Variation of Efficacy of Filtering Face Pieces Respirators Over Time in a Dental Setting: A Pilot Study." *Dentistry Journal* 9: 36.

657. Juang, P. S. C., and P. Tsai. 2020. "N95 Respirator Cleaning and Reuse Methods Proposed by the Inventor of the N95 Mask Material." *Journal of Emergency Medicine* 58: 817–820.

658. Claas, E. C. J., et al. 1998. "Human Influenza A H5N1 Virus Related to a Highly Pathogenic Avian Influenza Virus." *Lancet* 351: 472–477.

659. MacIntyre, C. R., and A. A. Chughtai. 2015. "Facemasks for the Prevention of Infection in Healthcare and Community Settings." *British Medical Journal* 350: h694.

660. MacIntyre, C. R., et al. 2009. "Face Mask Use and Control of Respiratory Virus Transmission in Households." *Emerging Infectious Diseases* 15: 233–241.

661. MacIntyre, C. R. 2025. Interview by author, July 14, 2025.

662. MacIntyre, C. R., et al. 2011. "A Cluster Randomized Clinical Trial Comparing Fit-Tested and Non-Fit-Tested N95 Respirators to Medical Masks to Prevent Respiratory Virus Infection in Health Care Workers." *Influenza and Other Respiratory Viruses* 5: 170–179.

663. Chopra, J., et al. 2021. "The Influence of Gender and Ethnicity on Facemasks and Respiratory Protective Equipment Fit: A Systematic Review and Meta-Analysis." *BMJ Global Health* 6.

664. Shaw, K. A., et al. 2021. "The Impact of Facemasks on Performance and Physiological Outcomes During Exercise: A Systematic Review and Meta-Analysis." *Applied Physiology, Nutrition, and Metabolism* 46: 693–703.

665. Sickbert-Bennett, E. E., et al. 2020. "Filtration Efficiency of Hospital Face Mask Alternatives Available for Use During the COVID-19 Pandemic." *JAMA Internal Medicine* 180: 1607–1612.

666. Greenhalgh, T., et al. 2024. "Masks and Respirators for Prevention of Respiratory Infections: A State of the Science Review." *Clinical Microbiology Reviews* 37: e0012423.

667. Jehn, M., et al. 2021. "Association Between K-12 School Mask Policies and School-Associated COVID-19 Outbreaks – Maricopa and Pima Counties, Arizona, July-August 2021." *Morbidity and Mortality Weekly Report* 70: 1372–1373.

668. Doyle, T., et al. 2021. "COVID-19 in Primary and Secondary School Settings During the First Semester of School Reopening – Florida, August-December 2020." *Morbidity and Mortality Weekly Report* 70: 437–441.

669. Gettings, J., et al. 2021. "Mask Use and Ventilation Improvements to Reduce COVID-19 Incidence in Elementary Schools – Georgia, November 16-December 11, 2020." *Morbidity and Mortality Weekly Report* 70: 779–784.

670. Nelson, S. B., et al. 2023. "Prevalence and Risk Factors for School-Associated Transmission of SARS-CoV-2." *JAMA Health Forum* 4: e232310.

671. Boulos, L., et al. 2023. "Effectiveness of Facemasks for Reducing Transmission of SARS-CoV-2: A Rapid Systematic Review." *Philosophical Transactions of the Royal Society A: Mathematical, Physical and Engineering Sciences* 381: 20230133.

672. Li, J., et al. 2020. "Epidemiology of COVID-19: A Systematic Review and Meta-Analysis of Clinical Characteristics, Risk Factors, and Outcomes." *Journal of Medical Virology* 93: 1449–1458.

673. O'Driscoll, M., et al. 2020. "Age-Specific Mortality and Immunity Patterns of SARS-CoV-2." *Nature* 590: 140–145.

674. Morawska, L. 2025. Interview by author, May 2, 2025.

675. Ibid.

676. Gottlieb, S. 2021. "*Lidia Morawska*." Time. Accessed June 3, 2025. https://time.com/collection/100-most-influential-people-2021/6095975/lidia-morawska/.

677. Morawska, L., et al. 2020. "How Can Airborne Transmission of COVID-19 Indoors Be Minimised?" *Environment International* 142: 105832.

678. Nightingale, F. 1863. *Notes on Hospitals*. London, UK: Longman, Green, Longman, Roberts and Green.

679. Bostridge, M. 2008. *Florence Nightingale: The Woman and Her Legend*. London, UK: Penguin Books.

680. Nightingale, F. 1863. *Notes on Hospitals*. London, UK: Longman, Green, Longman, Roberts and Green.

681. Ibid.

682. Campbell, M. 2005. "What Tuberculosis Did for Modernism: The Influence of a Curative Environment on Modernist Design and Architecture." *Medical History* 49: 463–488.

683. Andrewes, C. H. 1965. *The Common Cold*. London, UK: Weidenfeld and Nicolson.

684. Arundel, A. V., et al. 1986. "Indirect Health Effects of Relative Humidity in Indoor Environments." *Environmental Health Perspectives* 65: 351–361.

685. Hennessy, P. 2013. "*Eminent Victorians Dropped From History Curriculum in Gove U-Turn*." Telegraph.co.uk. Accessed May 30, 2025.

https://www.telegraph.co.uk/news/uknews/10136732/Eminent-Victorians-dropped-from-history-curriculum-in-Gove-U-turn.html.

686. UK Government and Parliament. 2014. "*Petition: Reinstate Florence Nightingale Into the National Curriculum for Schools.*" Accessed May 30, 2025. https://petition.parliament.uk/archived/petitions/52560.

687. Satish, U., et al. 2012. "Is CO2 an Indoor Pollutant? Direct Effects of Low-To-Moderate CO2 Concentrations on Human Decision-Making Performance." *Environmental Health Perspectives* 120: 1671–1677.

688. Gaihre, S., et al. 2014. "Classroom Carbon Dioxide Concentration, School Attendance, and Educational Attainment." *Journal of School Health* 84: 569–574.

689. Haverinen-Shaughnessy, U., et al. 2011. "Association Between Substandard Classroom Ventilation Rates and Students' Academic Achievement." *Indoor Air* 21: 121–131.

690. Wargocki, P., et al. 2020. "The Relationships Between Classroom Air Quality and Children's Performance in School." *Building and Environment* 173: 106749.

691. Morawska, L., et al. 2024. "Mandating Indoor Air Quality for Public Buildings." *Science* 383: 1418–1420.

692. Brown, C. J. F., and M. Mon-Williams. 2025. Interview by author, May 28, 2025.

693. Ibid.

694. Ibid.

695. Gouia, I., et al. 2024. "Epidemiology of Childhood Asthma in the UK." *Journal of Asthma and Allergy* 17: 1197–1205.

696. Seltzer, J. M., and M. J. Fedoruk. 2007. "Health Effects of Mold in Children." *Pediatric Clinics of North America* 54: 309–333.

697. Vitoratou, D.-I., et al. 2023. "Obesity as a Risk Factor for Severe Influenza Infection in Children and Adolescents: A Systematic Review and Meta-Analysis." *European Journal of Pediatrics* 182: 363–374.

698. Pal, S. 2025. Interview by author, June 2, 2025.

699. Brown, C. J. F., and M. Mon-Williams. 2025. Interview by author, May 28, 2025.

700. Ibid.

701. Xe Currency Converter. 2025. "*Historical Rate Tables.*" Newmarket, Canada: Xe Corporation Inc. Accessed September 5, 2025. https://www.xe.com/currencytables/?from=GBP&date=2025-07-01#table-section.

702. Philips. 2019. "*3000i Series.*" Amsterdam, Netherland: Koninklijke Philips NV. Accessed June 3, 2025. https://www.philips.co.uk/c-p/AC3033_30/3000i-series-air-purifier.

703. Morawska, L., et al. 2024. "Mandating Indoor Air Quality for Public Buildings." *Science* 383: 1418–1420.

704. Renna, S. 2025. Interview by author, June 16, 2025.

705. Xe Currency Converter. 2025. "*Historical Rate Tables.*" Newmarket, Canada: Xe Corporation Inc. Accessed November 20, 2025. https://www.xe.com/currencytables/?from=EUR&date=2025-07-01#table-section.

706. Bonan, J., et al. 2025. *The Effect of Air Purifiers in Schools: Working Paper 25-17 June 2025*. Washington DC, USA: Resources for the Future.

707. Brown, C. J. F., and M. Mon-Williams. 2025. Interview by author, May 28, 2025.

708. Federal Public Service Health Food Chain Safety and Environment. 2024. "*Legal Framework Regarding Indoor Air Quality.*" Brussels, Belgium: Federal Public Service Health, Food Chain Safety and Environment. Accessed June 3, 2025.

709. Brown, C. J. F., and M. Mon-Williams. 2025. Interview by author, May 28, 2025.

710. Andrewes, C. H. 1965. *The Common Cold*. London, UK: Weidenfeld and Nicolson.

711. Tyrrell, D., and M. Fielder. 2002. *Cold Wars: The Fight Against the Common Cold*. Oxford: Oxford University Press.

712. Dunn, P. M. 2000. "Dr William Buchan (1729-1805) and His Domestic Medicine." *Archives of Disease in Childhood: Fetal & Neonatal Edition* 83: F71–73.

713. Buchan, W. 1790. *Domestic Medicine: Or a Treatise on the Prevention and Cure of Diseases by Regimen and Simple Medicines with an Appendix Containing a Dispensatory foe the Use of Private Medicines. Eleventh Edition*. London, UK: Strathan & Cadell.

714. Collier, R. 1974. *The Plague of the Spanish Lady: The Influenza Pandemic of 1918-1919*. London, UK: Macmillan.

715. Lucy. 2025. Interview by author, January 24, 2025.

716. ONS. 2023. "*Prevalence of Ongoing Symptoms Following Coronavirus (COVID-19) Infection in the UK: 30 March 2023.*" London: Office for National Statistics. Accessed February 25, 2025. https://www.ons.gov.uk/peoplepopulationandcommunity/healthandsocialcare/conditionsanddiseases/bulletins/prevalenceofongoingsymptomsfollowingcoronaviruscovid19infectionintheuk/30march2023.

717. Al-Aly, Z., et al. 2024. "Long COVID Science, Research and Policy." *Nature Medicine* 30: 2148–2164.